Operations and Properties of Algebra

Irma M. Pellei

i

<u>Preface</u>

Having taught mathematics from grades seven through twelve over the past eleven years, I have seen what countless math teachers everywhere see. Having graded papers replete with conceptual errors time and again, I came to understand the nature of so many student misunderstandings: students at all levels of ability confuse addition with multiplication, addition with subtraction and basic multiplication with exponentiation.

When I was in elementary school in the 1970s, everything was slower. The basic four-function calculator that you can get today in a dollar store was in its nascent stages and prohibitively expensive. All computations were done manually and the need for students to develop rapidity in basic computations was indispensable. It seemed as though we spent endless weeks learning to multiply two-digit numbers by two-digit numbers, three-digit numbers by two-digit numbers, three-digit numbers by three-digit numbers – indeed, I don't remember exactly when it came to an end. But this much I know: by the time we were finished, it was absolutely impossible for a student to confuse the operation of multiplication with addition.

Not so today. Math curricula are much denser now to reflect the pace of life. Students are introduced to algebra as early as the third grade, before they have been given the chance to differentiate between the skills for adding and multiplying decimals. Everything at once is thrown at these young ones, and only the strongest succeed.

It is in the spirit of gratitude and honor to all of my teachers that I wrote this back-to-basics review book, which is intended to provide a review of elementary algebra or, in the alternative, to provide a rapid introductory study. It is ideal reading for students who must tackle weaknesses in algebra before moving on to the more challenging courses in intermediate and advanced algebra at the high school or college level. It is also ideal reading for parents of school-age children who want to help their children in algebra but currently lack the ability to do so. It can be used for self-study or for professional development. In fact, it can be used by everyone.

What is unique, new and timely about this text is the order of presentation of the material. Traditionally, the basic operations of addition, subtraction, multiplication and division are taught within the context of distinct topics such as integers, radicals, monomials and the like. The summary chart in Appendix A demonstrates this treatment vertically. In this text, the presentation is redirected from the vertical treatment illustrated in Appendix

A to the horizontal one: each of the four basic operations is emphasized as a distinct topic in separate chapters, and the manipulation of the various numerical and algebraic terms are developed within the context of each operation. By means of this set up, the student will be able to internalize the essence of each operation as it applies to the totality of algebra and differentiate the operation from the other three, providing for quicker advancement in algebraic skill. It is assumed that the student already has executed all four of the basic operations and will be prepared to apply, for example, subtraction, multiplication or division when they are needed to execute an addition problem. Such is the nature of the four operations; all are interdependently needed to effect one of them. Notwithstanding this necessity, the title of each chapter indicates the operation to be emphasized therein.

The student will gain little new insight if this book simply is read piecemeal as any other reference text. Instead, this book should be read from beginning to end or, at the very least, one chapter at a time in sequence, as the first four chapters explore the common links in addition, subtraction, multiplication and division of all things mathematical. Read this book as you would a novel, except that this is not fiction; it is the true story of algebra. Seen in this way, algebra no longer seems dry, cold and impersonal as algebraic terms and the operations affecting them take on an organic, living character.

A word of advice and encouragement to those who see no value in what may appear to be "mindless rote," prevalent in the first four chapters. Frankly, a bad name has been attached to a good thing. Every problem requires attention; practicing to the point of rote, therefore, is never mindless. The most effective Math teachers know that in order to master the underlying mechanics of arithmetic, a certain amount of "drill and kill" is the inevitable, necessary prerequisite. One must accept the truth that there is no difference between repetitive practice in mathematics as in repetitive practice of scales and arpeggios for the musician, the positions for the ballet dancer, the jump shot for the basketball player, the slapshot for the ice hockey player, save one thing: for the aspiring artist or athlete, the motivating force behind the repetition is the desire to become the next Joshua Bell, Mikhail Baryshnikov, Michael Jordan, or Wayne Gretzky. The passion to attain greatness, ever present in the aspiring artist or athlete, is almost always lacking in the math student, through no fault of his own. Recognizing the beauty in mathematics requires more effort because the beauty is subtle and not nearly as thrilling. Once the mechanics have been mastered, however, the world of abstract thinking and problem solving opens up; therein lies the thrill.

I. Pellei
Valley Stream, New York
July 2011

TABLE OF CONTENTS

ADDITION   1
    Signed Numbers   1
    Absolute Value   2
    Addition Rule # 1   4
    Addition Rule # 1 Explained   4
    Addition Rule # 2   5
    Addition Rule # 2 Explained   6
    Adding Integers   7
    Adding Numerical Fractions   8
    Adding Decimals   14
    Adding Numbers Expressed in Scientific Notation 16
    Adding Monomials   19
    Adding Binomials and Polynomials   25
    Adding Radicals   27
    Adding Algebraic Fractions   33
SUBTRACTION   44
    The Subtraction Rule   45
    Subtracting Integers   47
    Subtracting Numerical Fractions   48
    Subtracting Decimals   51
    Subtracting Numbers Expressed in Scientific Notation 52
    Subtracting Monomials   53
    Subtracting Binomials and Polynomials   55
    Subtracting Radicals   57
    Subtracting Algebraic Fractions   58
MULTIPLICATION   63
    Multiplication Rule # 1   66
    Multiplication Rule # 1 Explained   67
    Multiplication Rule # 2   67
    Multiplication Rule # 2 Explained   68
    Multiplying Integers   69
    Multiplying Numerical Fractions   69
    Multiplying Decimals   71
    Multiplying Numbers Expressed in Scientific Notation 72
    Multiplying Monomials   75
    Multiplying a Monomial by a Polynomial   79
    Multiplying Binomials   81
    Multiplying Polynomials   84
    Multiplying Radicals   85
    Multiplying Algebraic Fractions   88

DIVISION   92
    Division Rule # 1   94
    Division Rule # 2   94
    Division Rules # 1 and # 2 Explained   94
    Dividing Integers   94
    Dividing Numerical Fractions   95
    Dividing Decimals   97
    Dividing Numbers Expressed in Scientific Notation   99
    Dividing Monomials   100
    Factoring Binomials   104
    Dividing a Binomial by Monomial   108
    Factoring Trinomials   109
    Factoring and Dividing Radicals   115
    Factoring and Dividing Algebraic Fractions   117
ORDER OF OPERATIONS   119
    Evaluating Numerical Expressions   119
    Evaluating Algebraic Expressions   123
APPLICATIONS IN ONE VARIABLE:
    X AS AN UNKNOWN   125
    Solving Single Variable First Degree Equations
        in One Step   128
    Solving Single Variable First Degree Equations
        in Two Steps   131
    Solving Single Variable First Degree Equations
        in Three or More Steps   133
    Solving Single Variable First Degree Inequalities
        in One or More Steps   137
    Solving Single Variable Second Degree Equations   138
    Word Problems   140
    Consecutive Integer Problems   143
    Motion Problems   146
    Mixture Problems   150
    The Pythagorean Theorem   152
    Ratio and Proportion   154
    Percent   157
    Solving Single Variable Radical Equations   163
    Solving Fractional Equations   164
APPLICATIONS IN TWO VARIABLES:
    X AND Y AS CHANGING VALUES   166
    Expressing One Variable in Terms of the Other(s)   166
    The Coordinate Plane   168

The General Form of a Linear Equation   169
The Slope Formula   174
The Slope-Intercept Form of a Linear Equation   177
Graphing Lines of the Form y = b   180
Graphing Lines of the Form x = a   181
The Point-Slope Form of a Linear Equation   181
Writing the Equation of a Line Given the Slope
  and One Point   182
Writing the Equation of a Line Given Two Points   183
Solving Systems of Linear Equations in Two Variables   184
Graphing Systems of Linear Inequalities in Two Variables   186
Quadratic Equations and the Parabola   189
Completing the Square   195
The Quadratic Formula   197
Appendix A: Summary Chart of Operations of Algebra   201
Appendix B: Decimal Place Value and Scientific Notation   205
Appendix C: Answers   207

ADDITION

In algebra, addition is the first of four basic operations (subtraction, multiplication and division being the other three operations) and is often referred to by the phrase "combining like terms." This phrase eludes many students. Combining like terms should be thought of as taking an inventory of each type of term in the way retailers take inventory of the quantity that they have or need of different items. This chapter will develop which terms are like each other and which therefore can be consolidated or combined into a single term by addition. We will also see which types of terms are not initially like each other but can be changed in appearance so that they, too, may be combined into a single quantity by addition. Lastly we will see which types of terms cannot be changed into each other and therefore cannot be combined with each other, but rather are expressed together with other unlike terms in an addition chain.

Combining like algebraic terms will be explored later in this chapter. First, let us explore the concepts of signed numbers and absolute value, and then state the two addition rules which govern in algebra.

Mathematically, addition is ALWAYS and ONLY expressed by the symbol "+". There is no other way to indicate addition symbolically in algebra. Thus, English phrases such as "increased by," " the sum of" and "added to" are always replaced by the symbol "+" when translating word problems from English into algebraic sentences.

Signed Numbers

The sign on a number or other kind of term indicates whether the term is positive or negative. Positive terms are usually expressed without a plus sign in front of them and will be expressed as such in this text. Negative terms are ALWAYS expressed with a negative sign in front of them. The exception to this is in accountancy, where negative terms are expressed inside of parentheses, but that format will not be used here. However, it is useful to think like an accountant: think of a positive number as having something, and a negative number as owing something.

The number 3 indicates positive three and means having three of something. On a number line, it is three units to the right of zero.

The number – 3 indicates negative three and means owing three of something. On a number line, it is three units to the left of zero.

On a number line, everything from least to greatest is ordered from left to right. Given the value – 2 as an example, anything to the right is greater in value, and anything to the left is less in value.

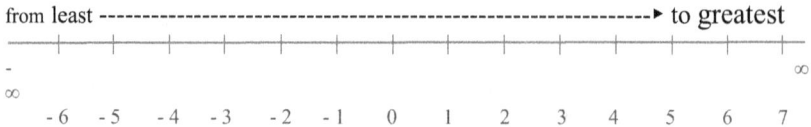

from least --→ to greatest

$-\infty$ ∞

-6 -5 -4 -3 -2 -1 0 1 2 3 4 5 6 7

Absolute Value

Now consider the concept of distance. When we measure distance, we express it as a positive quantity. In algebra, the absolute value of a number is a complicated-sounding phrase used to denote, simply, the distance or the number of units that number lies left or right of zero on a number line.

$$| x | = 3$$

↑

"the absolute value of x is 3"

where x represents, in this case, some expression whose value is either positive three or negative three. Vertical bars are used to denote the concept that the absolute value of the quantity inside is being sought before it has been evaluated; once the distance from zero has been determined, the vertical bars are removed and the absolute value of a number is ALWAYS expressed as a *positive* quantity.

From this, it follows that:

$\qquad | 3 | = 3$ ("the absolute value of positive three is three"),
and $\qquad | - 3 | = 3$ ("the absolute value of negative three is three"),
and $\qquad | 5 - 2 | = 3$ ("the absolute value of five minus two is three"),
and $\qquad | 2 - 5 | = | - 3 | = 3$ ("the absolute value of two minus five equals the absolute value of negative three, which equals three").

When one or more operations are expressed inside of the absolute value bars, those operations should be evaluated, following the order of operations, until a single numerical value is obtained. The last step is to express the resulting positive or negative number as a <u>positive number without the absolute value bars</u>. For purposes of this chapter, the student only need remember: *the absolute value of any real number is POSITIVE.*

We now turn to the two addition rules which govern the addition of all things mathematical. Each rule will be presented mathematically; immediately, an interpretation will follow.

The Vocabulary of Addition

Addend + Addend = Sum

<u>Addend</u> (n)– a term being added to another.
<u>Sum</u> (n)– the quantity that results from addition.

Addition Rule #1

When addends have the **SAME** SIGN (all are positive or all are negative), **ADD** their absolute values, and use the sign of the addends on the sum.

Positive Addends Have the Same Positive Sign and
are Added to Produce Positive Sums

$$3 + 5 = 8$$
$$1 + 9 = 10$$
$$17 + 24 = 41$$
$$23 + 45 + 56 = 124$$
$$.23 + .45 + .56 = 1.24$$

$$\frac{1}{7} + \frac{2}{7} + \frac{3}{7} = \frac{6}{7}$$

$$x + 2x + 3x = 6x$$

$$\sqrt{2} + 2\sqrt{2} + 3\sqrt{2} = 6\sqrt{2}$$

3

Negative Addends Have the Same Negative Sign and
are Added To Produce Negative Sums

$$^-3 + {}^-5 = {}^-\mathbf{8}$$
$$^-1 + {}^-9 = {}^-\mathbf{10}$$
$$- 17 + - 24 = {}^-\mathbf{41}$$
$$- 23 + - 45 + - 56 = {}^-\mathbf{124}$$
$$- .23 + - .45 + - .56 = -\mathbf{1.24}$$

$$\frac{-1}{7} + \frac{-2}{7} + \frac{-3}{7} = \frac{-\mathbf{6}}{\mathbf{7}}$$

$$^-x + {}^-2x + {}^-3x = {}^-\mathbf{6x}$$

$$^-\sqrt{2} + {}^-2\sqrt{2} + {}^-3\sqrt{2} = {}^-\mathbf{6\sqrt{2}}$$

Addition Rule #1 Explained

When applying this rule to any addition problem, think of money. Having money in one pocket or both is positive; owing money to one person or to two is negative. In both of these cases, you will be using Addition Rule # 1.

Ex. 1:   5   I have five dollars in my right pocket
            +   and
            __7__   I have seven dollars in my left pocket
            12   I have twelve dollars altogether

Ex. 2:   − 5   I owe five dollars to Mike
            +   and
            − 7   I owe seven dollars to Jessie
            − 12   I owe twelve dollars altogether

The rule is truly that simple: when adding positive quantities, add them and make the sum positive. When adding negative quantities, add them and make the sum negative. All that remains is to commit the rule to permanent, long-term memory. This, the first of two addition rules, governs the addition of all same-signed, like mathematical terms.

4

An all-too-common error is for students to express the sum of two negative addends as a positive sum. This confuses the rule for addition of negative addends with the rule for multiplication of negative factors.

Addition Rule # 2

When addends have **DIFFERENT** signs, **SUBTRACT** their absolute values and use the original sign of the term with the larger absolute value on the result.

Note that the second addition rule requires the use of subtraction. It would benefit the student to reread the previous sentence a few times until the concept is ingrained; *the second addition rule invokes the operation of subtraction.*

The Vocabulary of Subtraction

Minuend – Subtrahend = Difference

Minuend (n) – the first term in a subtraction operation.
Subtrahend (n) – the second term in a subtraction operation.
Difference (n) – the quantity that results from subtraction.

Procedure:

1. Take the absolute value of the addends.
2. Subtract: larger absolute value minus smaller absolute value.
3. Use the original sign of the minuend (larger absolute valued-term) on the result to complete your solution.

Ex. 3. **3 + – 7** Solution:
1. Take the absolute value of the addends.
$$| \ 3 \ | \ = 3$$
$$|-7| \ = 7$$
2. Subtract: $7 - 3 = 4$
= –4 3. Use the original sign of the minuend on the result to complete your solution.

5

Ex. 4. $-3 + 7$ <u>Solution</u>:

         1.   Take the absolute value of the addends.

$$|-3| = 3$$
$$|\ 7| = 7$$

         2.   Subtract: $7 - 3 = 4$

$= 4$ 3.   Use the original sign of the minuend on the result to complete your solution.

Addition Rule # 2 Explained

When you have money in one pocket (positive) and owe money to one person (negative), Addition Rule # 2 must be applied and subtraction must be performed. Assume in all cases that you take money from your pocket to pay the debt; after the debt is paid, the outcome depends on whether you have more money than you owe or still owe more after emptying your pocket.

Ex. 5:   3   I have three dollars
             +   and
         <u> − 7 </u>   <u>I owe seven dollars to Charlie</u>
         − 4   I still owe four dollars to Charlie

Ex. 6:   7   I have seven dollars
             +   and
         <u> − 3 </u>   <u>I owe three dollars to Alan</u>
         4   I still have four dollars

In the first example, negative seven, the minuend because it is farther from zero on the number line than three, is negative, so the outcome is negative. In the second example, seven, the minuend because it is farther from zero than negative three, is positive, so the outcome is positive.

Note the reason Addition Rule # 2 requires subtraction: there is a conflict between the positive and negative addend. The positive addend wants to move to the right of zero on the number line; the negative addend wants to move to the left of zero on the number line. Subtraction resolves the conflict without compromise; whatever remains after subtraction is the difference and whether it is negative or positive depends on which signed addend was farther from zero on the number line. Here we see the first instance of one operation's dependence on another; to effect Addition Rule # 2, addition is dependent upon subtraction.

Adding Integers

Now practice using the two rules with integers, which are (positive) counting numbers on the number line, their (negative) opposites and zero. Apply one of the two addition rules:

(1) If the signs are the **same, add** the absolute (positive) value of the addends and use the sign of both on the sum. (Positive plus positive is positive; negative plus negative is negative.)

(2) If the signs are **different, subtract** their absolute values and use the sign of the term farther from zero on the result.

1.
$$\begin{array}{r} 4 \\ + -6 \\ \hline \end{array}$$

2.
$$\begin{array}{r} -4 \\ +6 \\ \hline \end{array}$$

3.
$$\begin{array}{r} -4 \\ + -6 \\ \hline \end{array}$$

4.
$$\begin{array}{r} 3 \\ + -7 \\ \hline \end{array}$$

5.
$$\begin{array}{r} -3 \\ + 7 \\ \hline \end{array}$$

6.
$$\begin{array}{r} -3 \\ + -7 \\ \hline \end{array}$$

7.
$$\begin{array}{r} -8 \\ + -5 \\ \hline \end{array}$$

8.
$$\begin{array}{r} 8 \\ + -5 \\ \hline \end{array}$$

9.	$\begin{array}{r} -8 \\ +5 \\ \hline \end{array}$		14.	$\begin{array}{r} 11 \\ +-4 \\ \hline \end{array}$
10.	$\begin{array}{r} 2 \\ +-9 \\ \hline \end{array}$		15.	$\begin{array}{r} -11 \\ +-4 \\ \hline \end{array}$
11.	$\begin{array}{r} -2 \\ +9 \\ \hline \end{array}$		16.	$\begin{array}{r} -10 \\ +-2 \\ \hline \end{array}$
12.	$\begin{array}{r} -2 \\ +-9 \\ \hline \end{array}$		17.	$\begin{array}{r} -10 \\ +2 \\ \hline \end{array}$
13.	$\begin{array}{r} -11 \\ +4 \\ \hline \end{array}$		18.	$\begin{array}{r} 10 \\ +-2 \\ \hline \end{array}$

Upon completion of the exercises, use a calculator to check your solutions. If the concepts remain unclear, it is because old habits die hard. Reread the preceding section, memorize the rules, mimic the procedure, attempt the problems again and then move on to the next section. Clarity and understanding will be attained with patient practice.

Adding Numerical Fractions

Adding fractions requires one to apply the same rules of addition, that is:

(1) when the signs of like fractions are the **same** (both positive or both negative), **add** the numerators and **keep** the common denominator;
(2) when the signs of like fractions are **different** (one positive and one negative), **subtract** the numerators and **keep** the common denominator.

The Anatomy of a Fraction

division bar

$\dfrac{1}{2}$ ← numerator
   ← denominator

reads "one divided by two" or "one-half."

proper fraction - a fraction whose numerator is less than its denominator. A proper fraction lies between zero and one, or zero and negative one, on the number line.
improper fraction - a fraction whose numerator is greater than its denominator. An improper fraction is greater than one, or less than negative one, on the number line.

Fractions are like terms when they have the same denominator.

It is important to understand that *DENOMINATORS are NEVER ADDED* under *any* circumstance or operation in mathematics. One need only reason that one-half of an apple pie added to another one-half of an apple pie adds up to two-halves, or one whole pie and *not* to two-fourths of a pie (which is one-half).

$$\frac{1}{2} + \frac{1}{2} = \frac{2}{2} = 1$$

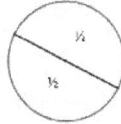

Notice that analogous equations can be written to express the fractions illustrated below. Also notice that the larger the denominator, the smaller the fraction, and the closer it is to zero on the number line.

$$\frac{1}{3} + \frac{1}{3} + \frac{1}{3} = \frac{3}{3}$$

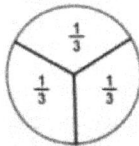

$$\frac{1}{4} + \frac{1}{4} + \frac{1}{4} + \frac{1}{4} = \frac{4}{4}$$

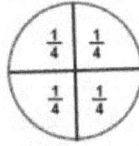

$$\frac{1}{5} + \frac{1}{5} + \frac{1}{5} + \frac{1}{5} + \frac{1}{5} = \frac{5}{5}$$

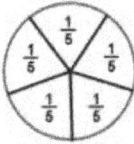

$$\frac{1}{7} + \frac{1}{7} + \frac{1}{7} + \frac{1}{7} + \frac{1}{7} + \frac{1}{7} + \frac{1}{7} = \frac{7}{7}$$

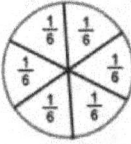

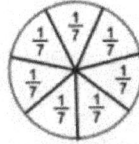

$$\frac{1}{6} + \frac{1}{6} + \frac{1}{6} + \frac{1}{6} + \frac{1}{6} + \frac{1}{6} = \frac{6}{6}$$

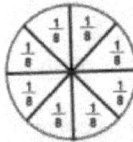

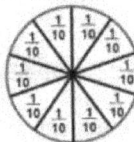

$$\frac{1}{8} + \frac{1}{8} + \frac{1}{8} + \frac{1}{8} + \frac{1}{8} + \frac{1}{8} + \frac{1}{8} + \frac{1}{8} = \frac{8}{8}$$

$$\frac{1}{10} + \frac{1}{10} + \frac{1}{10} + \frac{1}{10} + \frac{1}{10} + \frac{1}{10} + \frac{1}{10} + \frac{1}{10} + \frac{1}{10} + \frac{1}{10} = \frac{10}{10}$$

To understand the concept of a negative fraction, again the analogy of taking inventory, of completing an accountant's balance sheet, should be considered. A negative fraction should be thought of as a debt, lacking or owing of that fraction. Again, to add same-signed fractions, **add** the absolute (positive) value of the numerators, **keep** the common denominator and **keep** the sign of both addends on the sum. To add different-signed fractions, **subtract** the absolute value of the numerators, **keep** the common

9

denominator and use the sign of the numerator farther from zero on the difference.

Addition Rule # 1: **Fractions with Same Signs**	Addition Rule # 2: **Fractions with Different Signs**
$\dfrac{1}{5} + \dfrac{2}{5} = \dfrac{3}{5}$	$\dfrac{-1}{5} + \dfrac{3}{5} = \dfrac{2}{5}$
$\dfrac{-1}{5} + \dfrac{-2}{5} = \dfrac{-3}{5}$	$\dfrac{-2}{5} + \dfrac{4}{5} = \dfrac{2}{5}$
$\dfrac{2}{8} + \dfrac{4}{8} + \dfrac{7}{8} = \dfrac{13}{8}$	$\dfrac{-4}{5} + \dfrac{2}{5} = \dfrac{-2}{5}$
$\dfrac{-2}{8} + \dfrac{-4}{8} + \dfrac{-7}{8} = \dfrac{-13}{8}$	$\dfrac{-2}{8} + \dfrac{4}{8} + \dfrac{-7}{8} = \dfrac{-5}{8}$

It is rare for fractions to be given common denominators. Fractions usually have different denominators, representing different-sized pieces of a whole from a geometric perspective. Fractions with different denominators are UNLIKE TERMS and cannot be combined by addition UNTIL the fractions have been converted to like terms having common denominators. This brings us to the second instance of one operation's dependence upon other operations.

Converting fractions to like terms having common denominators involves multiplication of each fraction by another fraction that has a value of one.

The Multiplicative Identity Property

$a \cdot 1 = a$ any number or term times one equals itself.

From this it follows that

$\dfrac{a}{1} = a$ any number or term divided by one equals itself.

and

$\dfrac{a}{a} = 1$ any number divided by itself equals one.

The multiplicative identity property in its first and third usages is indispensable to converting unlike fractions into like terms. In the property's first usage, any term multiplied by the number one equals itself. In its third usage, the number one can be expressed as any fraction whose numerator and denominator are equal. To convert unlike fractions into like fractions, each fraction will be multiplied by a fraction equivalent to one, thereby giving the addends the same denominator. If the original fraction is negative, multiplying it by positive one will keep the new fraction negative. When the fractions have become like terms, they can be combined by addition. The following example illustrates these applications in the context of using Addition Rule # 1:

<u>Addend Fraction</u>   x One   =   <u>Fraction of Same Value</u>

$+$

$$\frac{1}{2} \quad x\ \frac{3}{3} \quad \begin{array}{l}\text{(multiply numerators)} = \\ \text{(multiply denominators)}\end{array} \quad \frac{3}{6}$$

$+$

$$\frac{1}{3} \quad x\ \frac{2}{2} \quad \begin{array}{l}\text{(multiply numerators)} = \\ \text{(multiply denominators)}\end{array} \quad \frac{2}{6}$$

$$\frac{5}{6}$$

(Add numerators / Keep common denominator)

The following example demonstrates the same procedure but now requires the application of Addition Rule # 2:

<u>Addend Fraction</u> x One   =   <u>Fraction of Same Value</u>

$+$

$$\frac{-3}{4} \quad x\ \frac{3}{3} \quad \begin{array}{l}\text{(multiply numerators)} = \\ \text{(multiply denominators)}\end{array} \quad \frac{-9}{12}$$

$+$

$$\frac{5}{6} \quad x\ \frac{2}{2} \quad \begin{array}{l}\text{(multiply numerators)} = \\ \text{(multiply denominators)}\end{array} \quad \frac{10}{12}$$

$$\frac{1}{12}$$

(Add numerators / Keep common denominator)

Sometimes the lowest common denominator is the product, or result from multiplication, of the two individual denominators. This is the easiest way to find a common denominator but may not always result in the least common denominator. In the previous example, four times six equals twenty-four but the least common multiple of four and six is twelve. The next example illustrates the simplicity of multiplying the two individual denominators to find the least common denominator. Again, Addition Rule # 2 applies.

Addend Fraction x One   =   Fraction of Same Value

$$\frac{-3}{5}$$ x $\frac{7}{7}$ (multiply numerators)   = $\frac{-21}{35}$

(multiply denominators)

+   +

$$\frac{2}{7}$$ x $\frac{5}{5}$ (multiply numerators)   = $\frac{10}{35}$

(multiply denominators)

$$\frac{-11}{35}$$

Add numerators

Keep common denominator

Adding fractions with unlike denominators requires skill in basic multiplication, skill in finding least common multiples, an understanding of the multiplicative identity property and skill in adding numerators with same and different signs. As with integers, recall that addition involves subtraction when the fractions have different signs. The dependence of addition on subtraction, multiplication and division (yes, the fraction equivalent to one is a division operation) is confusing to students who expect an operation to be a single-step instead of a multi-step procedure.

Now practice using Addition Rule # 1 and Addition Rule # 2.

Addend Fractions   x One   =   Fractions of Same Value

1. $\frac{3}{4}$

+ $\frac{2}{5}$

2. $\frac{-3}{4}$

+ $\frac{2}{5}$

3.
$$\frac{1}{4}$$

+

$$\frac{-1}{6}$$

4.
$$\frac{-1}{4}$$

+

$$\frac{-1}{6}$$

5.
$$\frac{-1}{2}$$

+

$$\frac{-1}{5}$$

6.
$$\frac{-3}{5}$$

+

$$\frac{-5}{4}$$

(Hint for Problem 7:   leave the second fraction as is, and change the first fraction to have the same denominator as the second.)

7.
$$\frac{9}{2}$$

+

$$\frac{-1}{4}$$

8.
$$\begin{array}{r} \underline{8} \\ 3 \end{array}$$
+
$$\begin{array}{r} \underline{-7} \\ 5 \end{array}$$

9.
$$\begin{array}{r} \underline{-4} \\ 9 \end{array}$$
+
$$\begin{array}{r} \underline{-5} \\ 6 \end{array}$$

10.
$$\begin{array}{r} \underline{4} \\ 3 \end{array}$$
+
$$\begin{array}{r} \underline{2} \\ 1 \end{array}$$

It would benefit the student at this point to absorb the key concepts of **ADD - KEEP** for the duration of this chapter, ever mindful that to **ADD** means **to add or to subtract**, depending upon whether the signs of terms are the same or different. This will aid the student in predicting the technique of adding all mathematical things to follow.

Adding Decimals

To add decimals, line up the decimal points and, using Addition Rules # 1 or # 2, **add** columns from right to left, carrying the tens place value of each column's sum to the top of the next column to the left. If necessary, fill in zeros to the end of a decimal to make addends the same length. Write in a decimal followed by any necessary zeros to the end of an integer. To decide whether to use Addition Rule # 1 or Addition Rule # 2 remember to consider whether a negative is in front of a number. **Keep** like place values (tenths, hundredths, thousandths, etc.) in alignment. Addition Rule # 1 is shown below:

```
                                      1        ← carry
       23.475                      23.475
+   14,375.6        becomes      + 14,375.600
                                  14,399.075
```

The quantity above can be thought of as fourteen thousand, three hundred ninety-nine dollars and seven-and-one-half cents. Bear in mind that each place value of a decimal is represented by some number from zero to nine. When ten is reached, that place value becomes zero and the place value to the left is increased by one; this can be seen in the tenths place of the previous example. Four plus six equals ten; zero is recorded below the line and the one is carried to the top of the column to the left. That one is added to three and five to equal nine. Now direct your attention to the last column on the right. Five-tenths equals one-half; hence .075 represents seven-and-one-half divided by one hundred. For a better understanding of decimal place value and how scientific notation is related to place value, see Appendix B.

Addition Rule # 2 requires subtraction of decimals with different signs. In order to effect subtraction, the student may need to rewrite the problem with the term farther from zero first, as in the following example:

```
                                  5 9 10   ← borrow from left
       23.475                   −  14,375.600
+ −  14,375.6       becomes     +        23.475    subtract
                               −  14,352.125
```

In the previous example, the six on the minuend is reduced to five so that zero to its right may become ten; this ten is reduced to nine so that the last zero may become ten, from which the last five in the subtrahend can be subtracted. The process of borrowing from the left requires going as far left as necessary in order to have a larger minuend from which to subtract.

If the student is ready to begin practicing now, however, try the following examples *without a calculator*, ever bearing in mind the two rules for addition, ever considering a positive quantity as an asset and a negative quantity as a debt. Do not be fooled by the order of the addends: you may have to rewrite the second addend to be first in order to subtract if the signs on the addends are different and the second addend is farther from zero.

```
1.                    3.75096
         +               10
```

2.   145.7396
      +   234.09817

3.   549,878.6348
      +   − 300.006

4.   − 754.58
      +   498.99

5.   − 2,096.25
      +   − 24,860.97

6.   − 45,934,086.8456
      +   12,000,070,482.03

7.   − 45,934,086.845
      +   − 12,000,070,482.03

Realize that in the above examples, the work of aligning decimal place value
has already been done for you in order to permit you to focus on whether to
use Addition Rule # 1 or Addition Rule # 2 and, if necessary, to rewrite the
second addend first to enable subtraction. The procedures in this section will
apply to all problems in the next section.

Adding Numbers Expressed in Scientific Notation

To add numbers expressed in scientific notation, **add** the coefficients, **keep**
the base ten and **keep** the exponent on the base ten. In scientific notation,
like terms are powers with the same base ten and the *same exponent* on the
base; therefore, they can be combined.

$$2.07 \times 10^8$$
$$+ \; 3.946 \times 10^8$$
$$6.016 \times 10^8$$

16

```
┌─────────────────────────────────────────────────────────────────┐
│                 The Anatomy of Scientific Notation                │
│                                                                   │
│                                        ✓ exponent                 │
│                    1.44      x     10⁵                            │
│                     ↗                ↑                            │
│                  Coefficient       base 10                        │
│                                                                   │
└─────────────────────────────────────────────────────────────────┘
```

coefficient- a numerical factor always placed first in a multiplication chain. In scientific notation, the convention is to express the coefficient as a decimal whose value is greater than or equal to one and less than ten.

base - the object of repeated multiplication to itself; the repeating factor.

exponent- indicates the number of times the base appears in the multiplication chain.

power - a term with a base and an exponent on the base. 10^5 reads "ten-to-the-fifth-power."

represents 1.44 x 10 x 10 x 10 x 10 x 10

or

144,000

If exponents on the base 10 are different, the terms are unlike and must be made like. The exponent of the addend of lower power must be **increased** to equal the exponent of the addend of higher power; move the decimal of its coefficient to the **left** by the same amount, then **add-keep-keep**.

$$2.07726 \times 10^{11} \text{ becomes}$$
$$+ \ 3.946 \quad \times 10^{8}$$

$$2.077260 \times 10^{11}$$
$$+ \ .003946 \times 10^{11}$$
$$\overline{2.081206 \times 10^{11}}$$

Note that the decimal of the second addend has been moved to the left three places and the exponent on the base ten has been increased by three. Also note that, because terms expressed in scientific notation are purely numerical, unlike terms can always be made like each other by making their powers equal.

Bear in mind that, as with all rules in mathematics, there is no exception to the requirement that unlike powers be made like before adding coefficients. Some interesting errors arise when 10^0 is one of the powers because 10^0

equals one. As anything times one equals itself, students will add the coefficients, ignore 10^0 and keep the other power, collapsing the correct procedure for adding numbers expressed in scientific notation with the rule for multiplying such numbers. Rather than reproduce the error here, let us remain focused on the correct procedure to which there is no exception and apply the skill to the following problems. As with adding decimals, the problems are to be done *without a calculator* in order to develop skill in manual computation and to fix the skill into permanent, long-term memory. First raise the lower power to the higher one, adjust the decimal of the coefficient to the left by the same amount and then see which term is farther from zero. If Addition Rule # 2 applies and the coefficient of the second addend is farther from zero, you may have to rewrite the problem to place the second addend first.

1. $\quad\quad\quad -5.9879 \times 10^4$
$\quad + \quad\quad -2.223 \;\; \times 10^4$

2. $\quad\quad\quad\quad 6.49329 \times 10^7$
$\quad + \quad -3.0927 \quad \times 10^8$

3. $\quad\quad\quad\quad 3.5926 \times 10^0$
$\quad + \quad -1.204 \;\; \times 10^3$

4. $\quad\quad\quad\quad 2.5847 \times 10^1$
$\quad + \quad\quad 1.9376 \times 10^0$

5. $\quad\quad\quad -1.09365 \times 10^5$
$\quad + \quad -2.94763 \times 10^2$

6. $\quad\quad\quad -4.9379 \;\; \times 10^2$
$\quad + \quad\quad\quad 7.70312 \times 10^0$

7. $\quad\quad\quad -6.4738 \times 10^2$
$\quad + \quad -3.8345 \;\; \times 10^3$

Adding Monomials

The Vocabulary of Multiplication

factor • factor = product

factor - a term multiplied to another.
product - the result of multiplication

18

$\dfrac{7\,x^{3}}{8}$ (one-eighth removed)

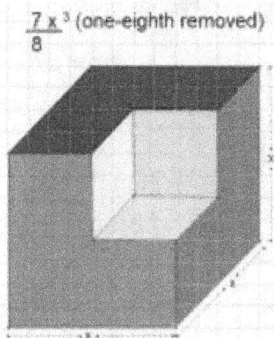

A monomial is an algebraic expression which may contain the operations of multiplication, division or exponentiation. A monomial never contains the operations of addition or subtraction; hence, it can be thought of as a multiplication chain. Examples of monomials follow:

19

$$7 \qquad -9 \qquad x \qquad 3x^2 \qquad \frac{4x^3}{3} \qquad {}^-y \qquad 5y^3$$

The first two monomials are simple integers with no variable. The third monomial above has no visible coefficient; an invisible coefficient of one is implied and the monomial represents one x. Negative y represents negative one y. The exponents on any monomial must be *whole numbers*, which are positive counting numbers or zero. The seven examples of monomials above are all **unlike terms** and therefore cannot be combined. **Monomials are like terms** when they have the **same powers**, that is, the same **variable base** and the **same exponent** on the base. Think of dollars, apples, apple pies, apple strudels, yams, yam pies and yam casserole. Apple pies have apples in them the way an x^2 term has x in it and yam casserole has yams in it the way a y^3 has y in it, but one would never say that one apple plus three apple pies equals four apple pies, or that one yam plus five yam casseroles equals six yams. Geometrically, x, x^2, x^3, y, y^2 and y^3 can be thought of as length, area and volume, all of which represent different measurements or different sized measurements, no one of which can be added to any other.

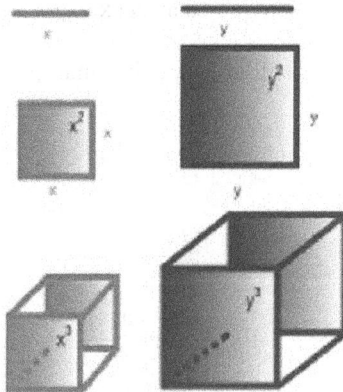

It should be obvious that x has an unknown value, and y has an unknown value different from x. It is worth repeating that no one of these terms can be combined with any other by means of addition. However, monomials are like terms when they have the same variable base and the same exponent on the base. An inventory of each type of term is taken – how many one has of each, how many one owes of each – by means of combining the term's coefficients with its attached sign.

20

Adding like monomials requires one to apply the same rules of addition, that is:

(1) when the signs of like monomials are the **same** (both positive or both negative), **add** the coefficients, **keep** the variable base and **keep** the exponent on the base;

(2) when the signs of like monomials are **different** (one positive and one negative), **subtract** the coefficients, **keep** the variable base and **keep** the exponent on the base.

Again the accountant's analogy should come to mind; just as one may have or owe money, one may have or owe a quantity of monomials. With this mind set, the student should apply the two rules of addition to combining like monomials, taking inventory of apples, apple pies, apple strudels, yams, yam pies and anything else with apples or yams in it.

It is critical to understand that unlike monomials with different dimensions can never be made like for purposes of addition. Addition can never be used to change length into area, length into volume or area into volume because these terms have different dimensions, natures or essences. A common error is to add the exponents as well as to add the coefficients, as in $3x^2 + 4x^2 \neq 7x^4$. This collapses the correct procedure with that for multiplying monomials. Adding exponents when adding monomials is incorrect because *adding exponents changes the dimension, nature or essence of the term*, which is not the goal of addition. The goal of addition and, as we will later see, subtraction, is to take inventory of **like terms**, not to change their nature. The correct procedure follows in the examples below.

Like Terms

$$x + - 3x + - 7x + 5x + 2x = {}^-2x$$

$$- 3x^2 + - 7x^2 + 5x^2 + 2x^2 = {}^- 3x^2$$

$$4y^3 + - 5y^3 + 7y^3 + - 2y^3 = 4y^3$$

$$.25x + - .7x + .23x + - .4x = - .62x$$

$$- 3x^2 + - .62x^2 + 1.04x^2 + 2.2x^2 = - .38x^2$$

$$5xy + - 7xy + - 3xy + xy = {}^- 4xy$$

Unlike Terms

Like terms are combined.
$$x + - 3x^2 + - 4x + 2x^2 + 5 + x^3$$
$$= x^3 + - x^2 + - 3x + 5$$
Unlike terms are linked together in an addition chain to express a complete inventory.

Like terms are combined.
$$5x^2 + - x + 6 + - 8x^2 + 4x + - 9$$
$$= {}^-3x^2 + 3x + {}^-3$$
Unlike terms are linked together in an addition chain to express a complete inventory.

Like terms are combined.
$$- 0.2x^2 + - x + - .4 + 9x^2 + .25x + - 9$$
$$= 8.8x^2 + - .75x + {}^-9.4$$
Unlike terms are linked together in an addition chain to express a complete inventory.

Note that in the fourth and fifth equation of the Like Terms examples, some coefficients are decimals, indicating some positive or negative pieces of a like power. *Decimals and integers are always like terms with respect to each other;* if they are the coefficients of like powers, these powers and pieces of powers must be combined for a complete and accurate inventory of the term they quantify.

In the sixth equation of the Like Terms examples, the like terms have two variables comprising a short multiplication chain: x and y both have an invisible exponent of one. The *xy* monomial is its own unique term and means *x* times *y*; geometrically, it represents a rectangle with length *x* and width *y*. All xy terms are like each other, and are unlike x^2, y^2, x^2y or xy^2. Thus, *we can expand the definition of like monomials* as follows: monomials are like terms when they have the same powers, or when they are composed of the same multiplication chain of like powers.

In the Unlike Terms examples, positive and negative x^2 terms are combined because they are like terms. Positive and negative x terms are likewise combined into a single term. Positive and negative integers are combined into a single term, and the three consolidated unlike terms are linked together in an addition chain, as if to state the inventory outcome as, "I owe three apple pies, I have three apples and I owe three dollars."

22

Monomials may have fractional coefficients indicating some positive or negative piece of a power. *If the powers are like terms, the monomials are like terms* and must be combined by addition for a complete and accurate inventory. In the case where fractional coefficients are unlike, they can and must be made like; then the like monomials can be combined using the procedure for **adding** fractional coefficients using Addition Rules # 1 and # 2, **keeping** the base and **keeping** the exponent on the base.

Like Terms
powers and coefficients are like

$$\tfrac{7}{8}x + -\tfrac{3}{8}x + -\tfrac{1}{8}x + \tfrac{5}{8}x = \tfrac{8}{8}\,x = 1x = \mathbf{x}$$

powers are like but coefficients are unlike

$$-3x^2 + \;-\tfrac{7}{8}x^2 + \tfrac{1}{4}x^2 + \tfrac{1}{2}x^2$$

coefficients are made like; powers are combined

$$= \;{}^{-24}/_8\,x^2 + -\tfrac{7}{8}\,x^2 + {}^2/_8\,x^2 + {}^4/_8\,x^2 \; = {}^{-25}/_8\,\mathbf{x^2}$$

powers are like but coefficients are unlike

$$4x^3 + -\tfrac{1}{2}x^3 + \tfrac{1}{4}x^3 + -\tfrac{1}{3}x^3$$

coefficients are made like; powers are combined

$$= 3\,{}^{12}/_{12}\,x^3 + -{}^6/_{12}\,x^3 + {}^3/_{12}\,x^3 + -{}^4/_{12}x^3 \; = \mathbf{3\,{}^5/_{12}\,x^3}$$

Unlike Terms

$$x + -3x^2 + -4x + 2x^2 + 5 + x^3$$
$$= \; x^3 + -x^2 + -3x + 5$$

$$5x^2 + -x + 6 + -8x^2 + 4x + -9$$
$$= \; {}^-3x^2 + 3x + {}^-3$$

Unlike coefficients of like terms are made like and combined

$$\tfrac{1}{10}x + -\tfrac{1}{4}x^3 + -\tfrac{1}{8}x^2 + -4x + \tfrac{2}{3}x^2 + 7 + \tfrac{1}{3}x^3$$
$$= {}^1/_{12}x^3 + {}^{13}/_{24}x^2 + -3{}^9/_{10}\,x + 7$$

Unlike terms are linked together in an addition chain to express a complete inventory

Unlike coefficients of like terms are made like and combined

$$\tfrac{3}{5}y + -\tfrac{1}{2}y^2 + -\tfrac{1}{10}y + 2y^2 + -2 + \tfrac{1}{12}y^3 \; = \; {}^1/_{12}y^3 + -{}^3/_2y^2 + -\tfrac{1}{2}y + -2$$

Unlike terms are linked together in an addition chain to express a complete inventory

Apply Addition Rules # 1 and # 2 now to the following problems.

1. $^-7x^2 + 5x^2$

2. $^-4c^2d + {}^-2c^2d$

3. $6xy^2 + 4xy^2 + {}^-9xy^2 + 2xy^2$

4. $8d + {}^-9d + {}^-3d + {}^-5d + d$

5. $^-9xy^2 + xy^2 + 2xy^2$

6. $.6m^2n + - .14m^2n + -.3\ m^2n$

7. $-.23a^2b + .41a^2b + - .53a^2b$

8. $\tfrac{3}{8}x^2y + - \tfrac{2}{3}x^2y + {}^-\tfrac{3}{4}x^2y$

9. $\tfrac{1}{3}xy^2 + \tfrac{5}{8}xy^2 + {}^-\tfrac{1}{2}xy^2$

10. $- \tfrac{3}{8}r^2s + \tfrac{2}{3}r^2s + - \tfrac{3}{4}r^2s + \tfrac{1}{2}r^2s$

11. $x + 11 + - 2x + - 4 + 3$

12. $0.5a + 8 + {}^-4 + {}^-2a$

13. $9x + 3 + {}^-11x + {}^-4 + {}^-1$

14. $^-\tfrac{1}{2}xy^2 + \tfrac{1}{4}x^2y + \tfrac{1}{3}xy^2 + \tfrac{2}{3}x^2y$

15. $-\tfrac{1}{4}a^2b + \tfrac{3}{4}ab^2 + - \tfrac{2}{3}a^2b + \tfrac{1}{3}ab^2$

16. $x^2 + 2x + 5 + 3x + 1$

17. $2x^2 + x + {}^-x^2 + 2x + 2$

18. $.3y^2 + .9y + {}^-.2y^2 + {}^-.3y + .8$

19. $\tfrac{2}{3}x^2 + \tfrac{1}{3}x + {}^-\tfrac{7}{8}x^2 + 2x + 2$

20. $y^2 + - \tfrac{1}{8}y + {}^-\tfrac{1}{4}y^2 + {}^-\tfrac{1}{2}y + - 3$

24

The solutions to the first ten problems are monomials. The solutions to problems 11 - 15 result in two unlike terms connected by addition, called **binomials**. The solutions to the last five problems result in three unlike terms connected by addition, called **trinomials**.

Adding Binomials and Polynomials

It is unusual for monomials to stand alone. Generally, unlike monomials are linked together in an addition chain. When they appear in an equation of two variables, they produce graphs of lines or curves in coordinate geometry. These equations will be explored in the chapter on applications in two variables. For now, it will suffice to review some basic vocabulary related to polynomials.

The Vocabulary of Polynomials

monomial (n) - a term, often composed of a multiplication chain containing one or more variables. The exponent on any variable MUST be a *whole number*, which is defined as a positive counting number or zero.

binomial (n) - two unlike monomials connected by addition or subtraction and forming an addition chain.

trinomial (n) - three unlike monomials connected by addition or subtraction and forming an addition chain.

polynomial (n) - one or more unlike monomials forming an addition chain.

first degree or **first order** (adj) - a polynomial whose monomial of highest power is one. When in an equation of two variables, produces the graph of a line. Also called **linear**.

second degree or **second order** (adj) - a polynomial whose monomial of highest power is two. When in an equation of two variables, produces the graph of a parabola. Also called **quadratic**.

nth **degree** or *nth* **order** (adj) - a polynomial whose monomial of highest power is *n*, where *n* is any positive integer.

This is where taking inventory gets interesting; increasingly complex, it requires the student's discerning eye. The polynomials already have been written so that like terms are in columns. Just apply the rule **add** coefficients- **keep** base - **keep** exponent on base; as always, to **add** means to add when Addition Rule #1 applies or to subtract when Addition Rule # 2 applies. When two or more variable powers are in a monomial, the exponents on the variables must match exactly; otherwise, the monomials are unlike.

1.
$$6x^2 + - 5$$
$$+ \quad 3x^2 + - 3$$

2.
$$3x^2 + 5$$
$$+ \quad 6x^2 + 3$$

3.
$$4x^2 + \quad 1$$
$$+ \quad 3x^2 + - 4$$

4.
$$- 5x^2 + 3x + - 8$$
$$+ \quad 2x^2 + - x + \quad 9$$

5.
$$12x^2 + - 4x + 4$$
$$+ \quad - 10x^2 + - 5x + 2$$

6.
$$- 7x^2 + - 9x + - 1$$
$$+ \quad - 3x^2 + \quad 5x + - 1$$

7.
$$10x^2 + 11x + 1$$
$$+ \quad - 5x^2 + - 6x + 1$$

8.
$$2x^2 + - 7x + \quad 5$$
$$+ \quad - 6x^2 + - 4x + - 2$$

9.
$$4x^2 + 10xy + \quad 9y^2$$
$$+ \quad 16x^2 + 32xy + 16y^2$$

10.
$$x^3 + \quad 3x^2y + \quad 3xy^2 + \quad y^3$$
$$+ \quad x^3 + ^-9x^2y + ^-9xy^2 + ^-27y^3$$

11.
$$x^4 + \quad 4x^3y + \quad 6x^2y^2 + \quad 4xy^3 + \quad y^4$$
$$+ \quad x^4 + ^-8x^3y + 24x^2y^2 + ^-32xy^3 + 16y^4$$

26

Notice that the work of aligning like terms in columns already has been done for you; this usually will not be the case. You may need to rewrite polynomials to align like monomials into columns in order to add. When you do so, momentarily ignore all signs and coefficients and focus strictly on the variables and the exponents attached to them. Once the polynomials have been reordered with like terms in columns, the addition is straightforward.

It is of highest importance for the student to realize that once a second or higher degree addition chain polynomial is formed, it becomes a unity requiring factoring or division in order to solve for x. This idea is so important that it bears repeating: *once a second or higher degree addition chain polynomial is formed, it becomes a unity requiring factoring or division in order to solve for x.* Like many chemical reactions, the formation of a polynomial by addition of unlike terms is irreversible. Adding or subtracting the like terms of two polynomials in order to form a new polynomial is frequently necessary, as was done in the preceding problems, but once the new unity is ready to be solved, it can only be accomplished through factoring or division. The ability to unlink the plus signs of linear polynomials in order to solve for x does not fall within this rule because linear polynomials are first degree.

A most common misconception is that the unlike monomials of a second or higher degree addition chain polynomial can be separated by simply unhinging the plus signs in the addition chain, by removing the addition links. Consider the following all-too-common errors made even at the level of calculus:

Error: $\dfrac{2x}{2x+5} = \dfrac{1}{5}$

Error: $\dfrac{2x}{x^2+5x+6} = \dfrac{A}{x^2} + \dfrac{B}{5x} + \dfrac{C}{6}$

Factoring and division of second order addition chain polynomials will be treated in depth in the chapter on division.

Adding Radicals

Most people like their numbers to look simple and neat, and math people are no different. Numbers that look neat are easier to manipulate. In this age of powerful technology, the need for neatness seems less compelling, but for centuries and as recently as forty years ago mathematics was developed with

27

this compelling need in the face of some very messy numbers. Hence, the radical. Before we add radicals, we will examine what a radical is.

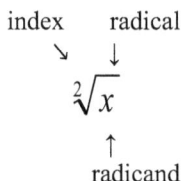

The Anatomy of a Radical

index   radical

$$\sqrt[2]{x}$$

radicand

Radical - the symbol for finding the root of a term.
Index - indicates the root to be taken of the term.
Radicand - the term whose root is to be taken.

Taking the square root or second root of a term is a process of thinking backwards: one must ask what number, when multiplied by itself, produces the radicand. Taking cube roots , fourth roots and higher *nth* roots requires thinking backwards as well, asking the student to consider what number is the repeating term in a multiplication chain of three, four or *n* links. In this text, we will deal with square roots only; higher roots are studied in higher level algebra courses. Some square roots are easy to calculate:

$\sqrt{4}$ = 2 ("the square root of four is two") (because 2 times 2 equals four)

$\sqrt{9}$ = 3 ("the square root of nine is three") (because 3 · 3 = 9)

It is critical to understand that taking the square root of a term is NOT division of the radicand by two. This all-too-common misconception arises because the square root of four is two and four divided by two is also two. But the square root of nine is three whereas nine divided by two is four-and one-half. The uniqueness of four as the sum and product of two repeated (2 + 2 = 4,   2 · 2 = 4) makes it an unfortunate place to begin teaching radicals; ironically, that is where one usually begins. Hence the misconception. Rather, taking the square root of a term is a process of thinking backwards in order to undo repeated multiplication of a term to itself; if it is to be thought of as division, then it is division of the radicand by its root in order to produce itself.

Taking the square root is the inverse operation of squaring a term; taking the cube root of a term is the inverse operation of cubing a term; taking the *nth* root of a term is the inverse operation of raising that term to the *nth* power, inasmuch as any term, once squared and square-rooted, or cubed and cube-rooted, becomes itself again.

On a number line, radical four lives at 2 on the number line because its value equals two; radical nine, at three, and so on.

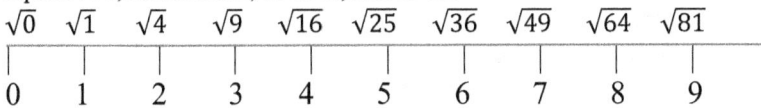

$\sqrt{0}$	$\sqrt{1}$	$\sqrt{4}$	$\sqrt{9}$	$\sqrt{16}$	$\sqrt{25}$	$\sqrt{36}$	$\sqrt{49}$	$\sqrt{64}$	$\sqrt{81}$
0	1	2	3	4	5	6	7	8	9

The radicals above are called **perfect squares** because they produce integers. At this point the student must commit to memory the following perfect squares:

$\sqrt{100} = 10$	$\sqrt{324} = 18$
$\sqrt{121} = 11$	$\sqrt{361} = 19$
$\sqrt{144} = 12$	$\sqrt{400} = 20$
$\sqrt{169} = 13$	$\sqrt{441} = 21$
$\sqrt{196} = 14$	$\sqrt{484} = 22$
$\sqrt{225} = 15$	$\sqrt{529} = 23$
$\sqrt{256} = 16$	$\sqrt{576} = 24$
$\sqrt{289} = 17$	$\sqrt{625} = 25$

But these are by no means the only radicals.

The square root of two, $\sqrt{2}$, is not a perfect square because no integer multiplied by itself gives two. We can see that radical two lives somewhere between radical one and radical four, or the integers one and two, on the number line; the question is where. After some guesswork, we can find that

1.4 x 1.4 = 1.96

This number is quite close to two, but we can get even closer:

1.41	x	1.41	=	1.9881
1.414	x	1.414	=	1.999396
1.4142	x	1.4142	=	1.99996164
1.41421	x	1.41421	=	1.999989924
1.414213	x	1.414213	=	1.999998409

In fact, we can get infinitely close. Clearly, a decimal with an infinite number of places after the decimal point is messy and still imprecise. Precision, a cornerstone of mathematics, is attained when a radical is left simply as it is. The square root of any number which is not a perfect square belongs to the category of *irrational number*.

This brings us to adding like radicals. An inventory of the quantity of radical twos or radical threes can be taken. As always, when no visible coefficient is present, the coefficient is presumed to be an invisible number one, or negative invisible one.

Like Terms

$$-\sqrt{3} + -5\sqrt{3} + 12\sqrt{3} = 6\sqrt{3}$$

$$\sqrt{5} + 5\sqrt{5} + -17\sqrt{5} = -11\sqrt{5}$$

$$\sqrt{7} + -2\sqrt{7} + 8\sqrt{7} = 7\sqrt{7}$$

$$x\sqrt{2} + 3x\sqrt{2} + -9x\sqrt{2} = -5x\sqrt{2}$$

$$x^2\sqrt{11} + 3x^2\sqrt{11} + -9x^2\sqrt{11} = -5x^2\sqrt{11}$$

$$xy\sqrt{6} + 3xy\sqrt{6} + -9xy\sqrt{6} = -5xy\sqrt{6}$$

Unlike Terms

$$13x\sqrt{2} + -9x\sqrt{3} + -11x\sqrt{2} + 4x\sqrt{3} + -x\sqrt{3} = 2x\sqrt{2} + -6x\sqrt{3}$$

$$-3\sqrt{5} + 10\sqrt{3} + -4\sqrt{3} + 7\sqrt{5} + -2\sqrt{3} + 14\sqrt{5} = 18\sqrt{5} + 4\sqrt{3}$$

$$23\sqrt{6} + -7\sqrt{2} + 3\sqrt{2} + -4\sqrt{6} + 21\sqrt{2} = 19\sqrt{6} + 17\sqrt{2}$$

It should be noted that the preferred order of terms in any multiplication chain is: sign - coefficient - radical - variable power. However, many students, through no fault of their own, tend to confuse when the radical ends and interpret any variable which may follow as being part of the radicand. It can be more confusing when a variable appears both inside and outside of a radical. In such cases, it is not entirely unconventional to express the variable or power in front of the radical.

It should begin to dawn on the student that when it comes to identifying like terms, the student must look at everything which follows the coefficient. Like terms will be identical in all of the following respects: variables, exponents on variables, radicands of radicals, and any multiplication chain of such terms.   When identifying like terms, the order of the links is unimportant because multiplication is commutative; all that matters is that every link be present. Indeed, the ONLY manner in which like terms may be different are the coefficients and the signs, positive or negative, attached to the front thereof.   The skeleton key of addition is: *like terms are multiplication chains which are identical link-for-link, regardless of their order, and except in their signs and coefficients.* By their signs and coefficients, inventory of like terms is taken to combine them into a single, distinct, unique monomial. These unique monomials are then linked to other unique but unlike monomials with plus signs in an addition chain to form a unity. This is how to add all things mathematical.

Now try some problems; only the first five have all like terms.

1. $2\sqrt{2} + -3\sqrt{2} + 5\sqrt{2} - 7\sqrt{2}$

2. $5\sqrt{3} + -4\sqrt{3} + -2\sqrt{3} + 7\sqrt{3}$

3. $\sqrt{5} + -4\sqrt{5} + -8\sqrt{5} + 12\sqrt{5}$

4. $x\sqrt{3x} + 2x\sqrt{3x} + -4x\sqrt{3x} + 9x\sqrt{3x}$

5. $4x^2\sqrt{x} + 3x^2\sqrt{x} + -6x^2\sqrt{x} + -13x^2\sqrt{x}$

6. $5\sqrt{3} + 2\sqrt{5} + -3\sqrt{3} + -7\sqrt{5}$

7. $-2\sqrt{2x} + 3x\sqrt{2x} + -3\sqrt{2x} + -5x\sqrt{2x}$

8. $-6x\sqrt{2} + x\sqrt{3} + -4x\sqrt{2} + -3x\sqrt{3} + 3\sqrt{2}$

9. $-9x\sqrt{3x} + 3\sqrt{3x} + -5x\sqrt{3x} + 4\sqrt{3x} + -12x\sqrt{3x} + 7\sqrt{2x}$

10. $8x^2\sqrt{5x} + -3x\sqrt{2x} + -4x^2\sqrt{5x} + -23x\sqrt{5x} + x^2\sqrt{5x} + -4x^2\sqrt{5x}$

When radicals are strictly numerical, often they are not written in simplest radical form. An example of this is

$$\sqrt{72}$$

In such cases the student must convert radicals into simplest form. Here we come to addition's dependence on division in order for it to be effected.

31

Converting radicals into simplest form involves pulling out a factor of the largest possible perfect square by taking its square root, leaving the smallest radicand under the radical. So

$$\sqrt{72} = \sqrt{36 \bullet 2} = 6\sqrt{2}$$

Realize that this technique can only be done when multiplication and division are the only operations under the radical. This will be explored in greater detail in the chapters on multiplication and division. The key to simplifying radicals is, by trial and error, to divide the radicand by 2, 3, 5, 6, or 7. One of these numbers will be the remaining radicand; the other number will be the largest possible perfect square whose square root can then be taken and converted into an integral factor, placed in front of the radical and creating a multiplication chain. Try dividing by two first; if the other number is not a perfect square, try to divide the radicand by three, and so on, until a perfect square is obtained. This procedure can be done only if the student is able to recognize a perfect square; if the student has not yet memorized them, a table of perfect squares must be used as a reference until they are memorized. Analyze the procedure in the following examples involving radicals which will become like terms, noting that the trial-and-error has not been shown:

$$\sqrt{75} + \sqrt{108} = \sqrt{25 \cdot 3} + \sqrt{36 \cdot 3} = 5\sqrt{3} + 6\sqrt{3} = 11\sqrt{3}$$

$$\sqrt{98} + \sqrt{18} = \sqrt{49 \cdot 2} + \sqrt{9 \cdot 2} = 7\sqrt{2} + 3\sqrt{2} = 10\sqrt{2}$$

$$\sqrt{125} + \sqrt{45} = \sqrt{25 \cdot 5} + \sqrt{9 \cdot 5} = 5\sqrt{5} + 3\sqrt{5} = 8\sqrt{5}$$

$$\sqrt{847} + \sqrt{567} = \sqrt{121 \bullet 7} + \sqrt{81 \bullet 7} = 11\sqrt{7} + 9\sqrt{7} = 20\sqrt{7}$$

Of course, it is possible that when simplifying radicals in an addition chain, unlike radicals will still result; nevertheless, it is preferable to simplify them:

$$\sqrt{500} + \sqrt{200} = \sqrt{100 \bullet 5} + \sqrt{100 \bullet 2} = 10\sqrt{5} + 10\sqrt{2}$$

On a few occasions one must divide the radicand by four as a last resort, because it will become the integer two; the other number, often prime, will remain as the radicand:

$$\sqrt{52} + \sqrt{60} = \sqrt{4 \bullet 13} + \sqrt{4 \bullet 15} = 2\sqrt{13} + 2\sqrt{15}$$

When pulling an integer from the radical, you must multiply it to any coefficient which may already be present in order to shorten the links in the multiplication chain and thereby to obtain an accurate inventory.

$$2\sqrt{75} + 3\sqrt{108}$$
$$= 2\sqrt{25 \bullet 3} + 3\sqrt{36 \bullet 3}$$
$$= 2 \bullet 5\sqrt{3} + 3 \bullet 6\sqrt{3}$$
$$= 10\sqrt{3} + 18\sqrt{3} = 28\sqrt{3}$$

$$5\sqrt{98} + 4\sqrt{18}$$
$$= 5\sqrt{49 \bullet 2} + 4\sqrt{9 \bullet 2}$$
$$= 5 \bullet 7\sqrt{2} + 4 \bullet 3\sqrt{2} = 10\sqrt{2}$$

Try these problems. The first six will result in the same radicand and all terms can then be combined. After the first six, combine whichever like terms you find and express your result of unlike radicals in an addition chain.

11. $\sqrt{8} + \sqrt{50}$

12. $3\sqrt{18} + 7\sqrt{72}$

13. $2\sqrt{27} + 5\sqrt{48} + 4\sqrt{147} + 6\sqrt{192}$

14. $\sqrt{75} + 8\sqrt{192} + \sqrt{432} + 11\sqrt{363}$

15. $\sqrt{80} + 2\sqrt{180} + 7\sqrt{245}$

16. $\sqrt{150} + \sqrt{384}$

17. $2x\sqrt{288} + \sqrt{338} + -3x\sqrt{72}$

18. $\sqrt{175} + -\sqrt{150} + -\sqrt{63}$

19. $-\sqrt{24} + x\sqrt{96} + \sqrt{20}$

20. $-\sqrt{24} + \sqrt{96} + \sqrt{20}$

Adding Algebraic Fractions

Algebraic fractions are fractions that have variables in either the numerator, the denominator or both. Examples of algebraic fractions follow:

33

$$\frac{-2}{x} \qquad \frac{x}{5} \qquad \frac{6}{x+3} \qquad \frac{x+3}{x-1} \qquad \frac{2x+4}{3x-7} \qquad \frac{x^2+5x+6}{x^2+7x+10}$$

Due to the fraction bar's having top priority in the order of operations, the entire numerator and the entire denominator must ALWAYS be treated as binomials or polynomials surrounded by parentheses, as though parentheses were protecting the expression. The fraction bar shares with parentheses the highest priority in the order of operations.

Order of Operations

The operations in any numerical or algebraic expression or equation must be evaluated in the following order:

1. Expressions involving operations inside **parentheses** are protected and must be evaluated first.   In a fraction, the **fraction bar** acts as protecting the entire numerator and the entire denominator with their respective sets of parentheses. **Absolute value bars** likewise act as parentheses protecting the entire expression therein. Expressions in all such operators are evaluated first.
2. **Exponents** and **radicals** are evaluated from left to right.
3. **Multiplication** and **division** are evaluated from left to right.
4. **Addition** and **subtraction** are evaluated from left to right.

Adding algebraic fractions requires one to apply the same rules of addition to the numerators that apply to combining the like terms of polynomials. The like terms of both numerators are combined using the two addition rules and the unlike sums are linked together in an addition chain. The like common denominator is kept.

As with numerical fractions, algebraic fractions are like terms when they have the same denominator.   An algebraic fraction contains at least one variable in the numerator, denominator, or both.

Like Terms

$$\frac{5}{x} + \frac{8}{x} = \frac{13}{x}$$

$$\frac{2}{x+2} + \frac{3}{x+2} = \frac{5}{x+2}$$

$$\frac{-6}{x-3} + \frac{-9}{x-3} = \frac{-15}{x-3}$$

$$\frac{12}{x+8} + \frac{-11}{x+8} = \frac{1}{x+8}$$

$$\frac{-17x}{x^2+5x+6} + \frac{x+6}{x^2+5x+6} = \frac{-16x+6}{x^2+5x+6}$$

Note that in the last example, the fractions are considered like terms because the denominators are like terms, even though the numerators are unlike. This does not present a problem, as the like terms of both numerators are combined and the unlike terms that result can be linked together in an addition chain to form a new numerator. As with numerical fractions, the size of the piece of a whole is determined by the denominator and only like-sized pieces may be added together. To imagine what any number or term divided by $x^2 + 5x + 6$ would look like, consider the diagram below, wherein the rectangle represents the whole; the number of pieces would be determined by an as-yet unknown value of x:

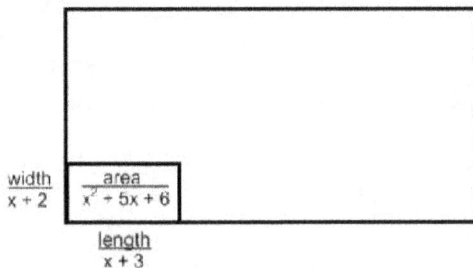

width
x + 2

area
$x^2 + 5x + 6$

length
x + 3

Although the image is abstract, one could make it more concrete by substituting the number one for x, making the length of the piece ¼ of the unknown length of the whole, the width of the piece ⅓ of the unknown width of the whole, the area of the piece $^1/_{12}$ of the unknown area of the whole, and the number of pieces 12, to equal one whole. Substituting different values of x would result in different-sized pieces. Likewise, one could substitute values for the length, width and area of the whole rectangle (say, 4, 2 and 8, respectively) which would work together to transform the model from an abstraction into a reality.

As with numerical fractions, algebraic fractions must be converted to have like denominators before their numerators can be added. The multiplicative identity property in its first and third usages (see page 10) is used to convert unlike algebraic fractions to like algebraic fractions having common denominators. The third usage of this property is extended to include any monomial, binomial or polynomial divided by itself to equal the number one, that is,

$$\frac{x}{x} = 1 \qquad \frac{x-3}{x-3} = 1 \qquad \frac{(x-3)(x-2)}{(x-3)(x-2)} = 1$$

From this, the examples below follow:

| Addend Fraction | x One | = | Fraction of Same Value |

$$\quad \frac{x}{2} \qquad\qquad x\ \frac{3}{3}\ \text{(multiply numerators)} \ =\ \frac{3x}{6}$$

$$+$$

$$\quad \frac{x}{3} \qquad\qquad x\ \frac{2}{2}\ \text{(multiply numerators)} \ =\ \frac{2x}{6}$$

(multiply denominators)

$$\frac{5x}{6}$$

Add numerators / Keep common denominator

| Addend Fraction | x One | = | Fraction of Same Value |

$$\quad \frac{-3}{4x} \qquad\qquad x\ \frac{3}{3}\ \text{(multiply numerators)} \ =\ \frac{-9}{12x}$$

$$+$$

$$\quad \frac{5}{6x} \qquad\qquad x\ \frac{2}{2}\ \text{(multiply numerators)} \ =\ \frac{10}{12x}$$

(multiply denominators)

$$\frac{1}{12x}$$

Add numerators / Keep common denominator

Note that in the next problem, nearly identical to the previous one, the first fraction lacks a factor of x in the denominator; the second fraction already has a factor of x in the denominator so it is not included in the fractional equivalent of one to which the second fraction is multiplied. Note also that adding the new numerators will require linking unlike terms in an addition chain.

36

Addend Fraction	x One	=	Fraction of Same Value	
$\dfrac{-3}{4}$	x $\dfrac{3x}{3x}$ (multiply numerators) (multiply denominators)	=	$\dfrac{-9x}{12x}$	Add numerators / Keep common denominator
+			+	
$\dfrac{5}{6x}$	x $\dfrac{2}{2}$ (multiply numerators) (multiply denominators)	=	$\dfrac{10}{12x}$ $\dfrac{-9x+10}{12x}$	

And yet one more variation follows: the lowest common denominator must now contain a factor of x • x, or x^2:

Addend Fraction	x One	=	Fraction of Same Value	
$\dfrac{-3}{4x}$	x $\dfrac{3x}{3x}$ (multiply numerators) (multiply denominators)	=	$\dfrac{-9x}{12x^2}$	Add numerators / Keep common denominator
+			+	
$\dfrac{5}{6x^2}$	x $\dfrac{2}{2}$ (multiply numerators) (multiply denominators)	=	$\dfrac{10}{12x^2}$ $\dfrac{-9x+10}{12x^2}$	

Again note that to add numerators of algebraic fractions means to combine any like terms using the two rules of addition and then to link the unlike terms in an addition chain. As with numerical fractions, denominators are NEVER added under any circumstance. Try some problems now.

<u>Addend Fractions</u>	<u>x One</u>	=	<u>Fractions of Same Value</u>

1. $\dfrac{3x}{4}$

$+$ $\dfrac{2x}{5}$

2. $\dfrac{-3x}{x+4}$

$+$ $\dfrac{2x}{x+4}$

3. $\dfrac{1}{4x}$

$+$

   $\dfrac{-1}{6x}$

4. $\dfrac{-1}{5x}$

$+$

   $\dfrac{-1}{8x}$

5. $\dfrac{-x}{x^2-5x+6}$

$+$

   $\dfrac{-x}{x^2-5x+6}$

38

Fractions	x One	=	Fractions of Same Value

6. $\dfrac{-3}{5x}$

$+$

$\dfrac{-5}{4x^2}$

_____

7. $\dfrac{-2}{3x}$

$+$

$\dfrac{-7}{4x}$

_____

8. $\dfrac{8}{3x}$

$+$

$\dfrac{-7}{5x^2}$

_____

9. $\dfrac{-4}{9}$

$+$

$\dfrac{-5}{6x^2}$

_____

10. $\dfrac{4}{3x^2}$

$+$

$\dfrac{2}{x}$

_____

Adding more complex unlike algebraic fractions requires the application of the distributive property.

The Distributive Property of Multiplication over Addition

$a(b + c) = ab + ac$ a factor being multiplied to an addition chain is multiplied to each and every term, like or unlike, in the chain.

From this it follows that

$a(b + c - d) = ab + ac - ad$ every term in an addition chain of any length is multiplied to the factor outside the parentheses.

While it is possible for the terms b and c to be like, very often the terms are unlike. The nature of the distributive property will be examined in greater depth in the chapter on multiplication. For now, it suffices to become proficient in the mechanics of the operation, as follows: when a term, a, appears in front of parentheses with no operation symbol in between the term and the parentheses, and inside parentheses there are terms connected by addition or subtraction, then the term outside the parentheses is multiplied to each and every term in the parentheses. If the term outside the parentheses is positive, the addition or subtraction operations are preserved in between the new products. If the term outside the parentheses is negative, the addition or subtraction operations are reversed.

Using the distributive property in the denominator is unnecessary; it is in the numerator of new equivalent fractions where the use is necessary. Consider

$$\frac{-6}{x-3} + \frac{-9}{x+2}$$

The least common denominator is the product of the two binomial denominators, that is, $(x - 3)(x + 2)$. It will be unnecessary for now to complete this multiplication to form a trinomial; multiplying binomials is reserved for the chapter on multiplication. However, the distributive property must be used in the numerators:

Addend Fractions	x One	=	Fractions of Same Value
$\dfrac{-6}{x-3}$	$\dfrac{(x+2)}{(x+2)}$	=	$\dfrac{-6(x+2)}{(x-3)(x+2)}$
+			+
$\dfrac{-9}{x+2}$	$\dfrac{(x-3)}{(x-3)}$	=	$\dfrac{-9(x-3)}{(x-3)(x+2)}$

It is fine to leave the denominators as they are, but the numerators must be expanded using the distributive property in order to combine like terms:

$$\frac{-6(x+2)}{(x-3)(x+2)} + \frac{-9(x-3)}{(x-3)(x+2)} = \frac{-6x+(-6\cdot2)+-9x+(-9\cdot-3)}{(x-3)(x+2)}$$

$$= \frac{-6x+-12+-9x+27}{(x-3)(x+2)} = \frac{-15x+15}{(x-3)(x+2)}$$

Note that when negative nine is multiplied to negative three, positive twenty-seven results. After using the distributive property, like terms are combined: twenty-seven plus negative twelve is fifteen, and negative 6x plus negative 9x is negative 15x. Try some problems now.

Fractions	x One	=	Fractions of Same Value

11.
$$\frac{-5}{x}$$
+
$$\frac{-2}{x-3}$$

[Hint: the common denominator will be x(x – 3)]

12.
$$\frac{7}{x+8}$$
+
$$\frac{-5}{x-6}$$

41

Fractions	x One	=	Fractions of Same Value

13.
$$\frac{6}{x+5}$$
$+$
$$\frac{-8}{x}$$
_____

14.
$$\frac{5}{x+11}$$
$+$
$$\frac{-2}{x}$$
_____

15.
$$\frac{-3}{x-7}$$
$+$
$$\frac{8}{3x}$$
_____

16.
$$\frac{-4}{5x^2}$$
$+$
$$\frac{13}{x-5}$$
_____

42

Fractions	x One	=	Fractions of Same Value

17. $\dfrac{-3}{x+1}$

$+$

$\dfrac{8}{x+7}$

18. $\dfrac{-2}{x-7}$

$+$

$\dfrac{8}{x-6}$

19. $\dfrac{-5}{x-2}$

$+$

$\dfrac{8}{x-5}$

20. $\dfrac{-4}{x^2+1}$

$+$

$\dfrac{3}{x-4}$

SUBTRACTION

Subtraction is the inverse operation of addition. Whereas to add at an elementary level means to increase by a certain amount, to subtract at an elementary level means to decrease by a certain amount. When any quantity is added to a mathematical term, subtraction of that same quantity from the sum results in the original term. So,

$$3 + 4 = 7, \qquad 7 - 4 = 3$$

We have explored addition at length, however, and have seen that to add means to add or to subtract, depending upon whether terms are same-signed or different in sign. Addition of signed terms no longer only means to increase in amount; sometimes it means to decrease in amount. What, then, is the nature of subtraction in the context of signed numbers and terms? The answer, simply, is this: **to subtract means to add the subtrahend's signed opposite,** called the **additive inverse**.

The Additive Inverse Property

$a + -a = 0$ any number or term plus its additive inverse equals zero.

The Additive Identity Property

$a + 0 = a$ any number or term plus zero, the additive identity element, equals itself.

From these it follows that

$b + a + -a = b$ any term, plus a second term, plus the additive inverse of the second term, equals the first term. [The first term plus zero equals itself.]

The additive inverse and additive identity properties are essential to advance the concept that subtraction is addition of the opposite. If subtraction is the inverse operation of addition, and if to add a term's additive inverse instead of the original term is to move the same distance in the opposite direction, then performing the inverse of the inverse preserves the operation.

44

The Subtraction Rule

In simpler terms, **to subtract means to add the signed opposite.**

Subtracting	=	Adding the Opposite
$7 - 3 = 4$		$7 + - 3 = 4$
$8 - 6 = 2$		$8 + - 6 = 2$
$10 - 3 = 7$		$10 + - 3 = 7$
$12 - 4 = 8$		$12 + - 4 = 8$

Bear in mind that every subtrahend in the first column is positive, whereas every subtrahend in the second column is negative. The key to every subtraction operation is to change it to an addition problem and to change, to reverse, the sign on the subtrahend. Reread the previous sentence to absorb the idea of subtraction's total dependence on addition: *the key to every subtraction operation is to change it to an addition problem and to change, to reverse, the sign on the subtrahend.* This singular rule of subtraction has no exception and can be summarized as follows:

- **keep** the sign on the minuend
- **plus**: change the subtraction symbol (–) to the addition symbol (+)
- change the sign of the subtrahend to the **opposite**. Once the problem has become an equivalent addition problem, select Addition Rule # 1 or Addition Rule # 2.

Consider the following examples:

Ex. 1.　　$6 - 8$　　"I must give Lucy eight dollars from my six."

becomes　　$6 + {}^- 8 = -2$

which now enjoys the analogy "I have six dollars and owe Lucy eight; therefore, I still owe Lucy two dollars."

Ex. 2.　　　$-6 - {}^- 8$　　　"I owe Lorraine six dollars and my other debt of eight dollars has been taken away"

becomes　$-6 + {}^+ 8 = 2$

analogous to, "I owe Lorraine six dollars and have eight; I still have two dollars."

45

In this example, the analogies are equivalent and make perfect sense to an accountant or the IRS. In contrast, subtraction of a negative term may be the single most overlooked operation by algebra students. The pervasive common error is to subtract a positive term instead of performing the double negative. It would benefit the student to memorize the idea that *subtracting a negative means adding a positive*, and to apply it in all instances throughout this chapter without exception. Students will be reminded to do so in each subchapter.

Ex. 3. $-8-6$ "I wrote Annette a check for six dollars, to be paid from my negative 8 dollar balance in the account."

becomes $-8 + -6 = -14$

interpreted as "I owe Annette six dollars and my account has a negative eight dollar balance; therefore I am fourteen dollars in debt altogether."

Rather than try to understand all of the analogies, it is better not to think when applying the rule for subtraction. Subtracting negatives and larger positives is not as intuitive as adding assets and debts to calculate one's net worth. It is better to *not* think, apply the **keep – plus – opposite** rule and *then* to think: Addition Rule # 1 or Addition Rule # 2. Oddly enough, once the process becomes second nature, the analogies all make sense.

Ex. 4. $^-8 - ^-6 = ^-8 + ^+6 = -2$

Ex. 5. $8 - 6 = 8 + ^-6 = 2$

Ex. 6. $8 - ^-6 = 8 + ^+6 = 14$

Ex. 7. $^-8 - 6 = ^-8 + ^-6 = ^-14$

Consider a few more ideas about subtraction before moving on to applications. Subtraction is not commutative.

The Commutative Property	
$a + b = b + a$	The order of addends does not affect the sum.
$ab = ba$	The order of factors does not affect the product.

As we saw in the first example, $6 - 8 = -2$ whereas $8 - 6 = 2$. Clearly, the order of minuend and subtrahend matters in subtraction. Reverse the minuend and subtrahend of any original problem and the new difference is the additive inverse of the original difference.

However, there are times when we want to facilitate some algebraic manipulation without, of course, violating the order of operations. In such a case, the way to achieve commutativity in an addition/subtraction chain of positive and negative terms is to change the subtraction operation to the equivalent addition operation. Thus

$$6 - 8 = 6 + {}^-8 = {}^-8 + 6 = {}^-2$$

This holds true for any subtraction operation, and leads us to one more idea about subtraction: **in every subtraction operation, the subtraction operation, the minus sign, is attached to the term that follows it, whether that term is positive or negative.** Note how that is the case for Example 1 just revisited. The minus sign is attached to eight, the subtrahend. In example 2, the minus sign is attached to negative eight, making it positive eight. Hence, every subtraction problem is really an addition problem in disguise. As the student becomes comfortable with this idea, the preference of using subtraction of positive subtrahends over addition of negatives as a means of minimizing signs and symbols becomes obvious.

One final word about subtraction: in algebraic word problems, the phrases "decreased by," "subtracted from," "less," "less than" all indicate subtraction.

Before moving on to solving problems, the student should have completed the entire chapter on addition first. The unique structure of this book is designed to set a thorough foundation for each distinct operation. Since successful subtraction is entirely dependent on solid addition skills, the student's progress through this chapter will accelerate only if she has worked through the first 43 pages. Make a fruitful time investment now.

Subtracting Integers

Now practice the rule **keep-plus-opposite.** Remember, do NOT think about the changes; just make them. Subtracting a positive means adding a negative, so turn the minus sign into plus and put a negative sign in front of a subtrahend that has no sign in front. Subtracting a negative means adding a positive, so make the double- minus a double-plus. Then look at your transformed addition problem and think about choosing between Addition

Rule # 1 or Addition Rule # 2. It may help to convert all the problems to addition first, as if you were working on an assembly-line, and then to go back to solve them. This technique can help to internalize the **keep-plus-opposite** procedure as an automated response any time subtraction is involved.

1.	$\begin{array}{r} 4 \\ -\ ^-6 \\ \hline \end{array}$		11.	$\begin{array}{r} 2 \\ -\ -9 \\ \hline \end{array}$
2.	$\begin{array}{r} ^-4 \\ -\ ^-6 \\ \hline \end{array}$		12.	$\begin{array}{r} -2 \\ -\ 9 \\ \hline \end{array}$
3.	$\begin{array}{r} 4 \\ -\ 6 \\ \hline \end{array}$		13.	$\begin{array}{r} 2 \\ -\ 9 \\ \hline \end{array}$
4.	$\begin{array}{r} -4 \\ -\ 6 \\ \hline \end{array}$		14.	$\begin{array}{r} -2 \\ -\ -9 \\ \hline \end{array}$
5.	$\begin{array}{r} 3 \\ -\ -7 \\ \hline \end{array}$		15.	$\begin{array}{r} -11 \\ -\ 4 \\ \hline \end{array}$
6.	$\begin{array}{r} -3 \\ -\ 7 \\ \hline \end{array}$		16.	$\begin{array}{r} 11 \\ -\ -4 \\ \hline \end{array}$
7.	$\begin{array}{r} -3 \\ -\ -7 \\ \hline \end{array}$		17.	$\begin{array}{r} -11 \\ -\ -4 \\ \hline \end{array}$
8.	$\begin{array}{r} -8 \\ -\ -5 \\ \hline \end{array}$		18.	$\begin{array}{r} -10 \\ -\ -2 \\ \hline \end{array}$
9.	$\begin{array}{r} 8 \\ -\ -5 \\ \hline \end{array}$		19.	$\begin{array}{r} -10 \\ -\ 2 \\ \hline \end{array}$
10.	$\begin{array}{r} -8 \\ -\ 5 \\ \hline \end{array}$		20.	$\begin{array}{r} 10 \\ -\ -2 \\ \hline \end{array}$

Subtracting Numerical Fractions

Everything you need to know about subtracting fractions, you already know. To review, subtraction of fractions is not commutative until after it becomes addition by way of **keep - plus - opposite**, so do this first. As with integers,

48

if you are subtracting a positive fraction, it becomes adding a negative fraction; subtracting a negative fraction becomes adding a positive fraction. Place the new sign in the subtrahend's numerator.   Once subtraction has become addition of the subtrahend's signed opposite, which is always the second fraction, the fractions must be given common denominators using the same procedure for adding fractions with unlike denominators; here note subtraction's dependence on multiplication in order to be effected.   After the fractions have become like terms, use Addition Rule # 1 or Addition Rule # 2. If you require a more thorough review, adding fractions begins on page 8.

Fractions				x One Value	=	Fractions of Same
$\dfrac{3}{4}$	=	$\dfrac{3}{4}$	x	$\dfrac{5}{5}$	=	$\dfrac{15}{20}$
$-$						
$\dfrac{2}{5}$	= $+$	$\dfrac{-2}{5}$	x	$\dfrac{4}{4}$	$+$	$\dfrac{-8}{20}$
						$\dfrac{7}{20}$

Now try the following problems.

Fractions	x One	=	Fractions of Same Value

1. $\dfrac{3}{4}$

  $-$ $\dfrac{2}{5}$

2. $\dfrac{-3}{4}$

  $-$ $\dfrac{2}{5}$

Fractions	x One	=	Fractions of Same Value

3. $\dfrac{1}{4}$

$-$

$\dfrac{-1}{6}$

4. $\dfrac{-1}{4}$

$-$

$\dfrac{-1}{6}$

5. $\dfrac{-1}{2}$

$-$

$\dfrac{-1}{5}$

6. $\dfrac{-3}{5}$

$-$

$\dfrac{-5}{4}$

7. $\dfrac{9}{2}$

$-$

$\dfrac{-1}{4}$

(Hint: leave this fraction as is, and change the first fraction to have the same denominator as this one.)

Fractions	x One	=	Fractions of Same Value

8. $\dfrac{8}{3}$

−

   $\dfrac{-7}{5}$

9. $\dfrac{-4}{9}$

−

   $\dfrac{-5}{6}$

10. $\dfrac{4}{3}$

−

   $\dfrac{2}{1}$

Subtracting Decimals

Again, **keep-plus-opposite** is performed first. Subtracting a positive means adding a negative, so turn the minus sign into plus and put a negative sign in front of a subtrahend that has no sign in front. Subtracting a negative means adding a positive, so make the double- minus a double-plus. You may need to rewrite problems requiring Addition Rule # 2 so that the term farther from zero is above. Add a decimal and as many zeros as required to perform Addition Rule # 2 on the transformed addition problem. For a more thorough review, adding decimals begins on page 14.

Ex:

$$
\begin{array}{r}
3.75096 = \\
- \quad 10
\end{array}
\qquad
\begin{array}{r}
3.75096 = \\
+ \; -10
\end{array}
\qquad
\begin{array}{r}
-10.00000 \\
+ \quad 3.75096 \\
\hline
-6.24904
\end{array}
$$

51

1.   145.7396
    −   234.09817

2.   549,878.6304827
    −   − 300.0006

3.   − 754.58
    −   498.99

4.   − 2,096.25
    −   − 24,860.97

5.   − 45,934,086.84561
    −   12,000,070,482.03

6.   − 45,934,086.84561
    −   − 12,000,070,482.03

7.   − 459,872,940.093845
    −   − 2,637,930.095298

8.   54,983,093.90980867
    −   − 123,947,832.0987672634

Subtracting Numbers Expressed in Scientific Notation

As always, first perform **keep-plus-opposite** to transform the problem to addition. Subtracting a positive means adding a negative, so turn the minus sign into plus and put a negative sign in front of a subtrahend that has no sign in front. Subtracting a negative means adding a positive, so make the double- minus a double-plus. Then follow the procedure for adding numbers expressed in scientific notation beginning on page 16.

Ex:

$$-5.9879 \times 10^3 = \quad -5.9879 \times 10^3 = \quad 2.22300 \times 10^4$$
$$-\quad -2.223 \times 10^4 \quad + \quad +2.223 \times 10^4 \quad +-0.59879 \times 10^4$$
$$\overline{1.62421 \times 10^4}$$

1. 6.49329×10^7
   $-$ $-3.0927 \;\; \times 10^8$

2. 3.5926×10^0
   $-$ $-1.204 \;\; \times 10^3$

3. 2.5847×10^1
   $-$ $1.9376 \times \; 10^0$

4. -1.09365×10^5
   $-$ -2.94763×10^2

5. $-4.9379 \;\; \times 10^2$
   $-$ 7.70312×10^0

6. -6.4738×10^2
   $-$ $-3.8345 \;\; \times 10^3$

7. 1.084738763×10^2
   $-$ 2.056346389×10^5

8. 1.36822534×10^6
   $-$ $-7.94678356 \;\; \times 10^4$

9. 1.36822534×10^6
   $-$ -7.94678356×10^8

Subtracting Monomials

The subtraction chains in the problems below will be transformed into addition chains using **keep-plus-opposite-plus-opposite-plus-opposite** for as many links as there are in the chain. Only the first monomial's sign will be kept. Then apply Addition Rules # 1 or # 2:

(1) when the signs of like monomials are the **same** (both positive or both negative), **add** the coefficients, **keep** the variable base and **keep** the exponent on the base;

(2) when the signs of like monomials are **different** (one positive and one negative), **subtract** the coefficients, **keep** the variable base and **keep** the exponent on the base.

53

As with addition of monomials, only the first seven problems contain all like terms ready to combine. Problems 8 through 10 contain like terms with unlike fractional coefficients which must first be made like before the terms can be combined. Problems 11 through 13 contain two different types of like terms; problems 14 and 15 contain two types of like terms with unlike fractional coefficients. Problems 16 through 18 contain three different kinds of like terms. Problems 19 and 20, designed to challenge the student, contain three different kinds of like terms, complicated by unlike fractional coefficients. Adding monomials can be reviewed beginning on page 18.

Ex. $\quad 6x^2 - 4x - {}^-2x^2 - 7x - 2 = 6x^2 + {}^-4x + {}^+2x^2 + {}^-7x + {}^-2$
$$= 8x^2 - 11x - 2$$

1. $\quad {}^-7x^2 - 5x^2$

2. $\quad {}^-4c^2d - {}^-2c^2d$

3. $\quad 6xy^2 - 4xy^2 - {}^-9xy^2 - 2xy^2$

4. $\quad 8d - {}^-9d - {}^-3d - {}^-5d - d$

5. $\quad {}^-9xy^2 - xy^2 - 2xy^2$

6. $\quad .6m^2n - -.14m^2n - -.3m^2n$

7. $\quad -.23a^2b - .41a^2b - -.53a^2b$

8. $\quad \tfrac{3}{8}x^2y - -\tfrac{2}{3}x^2y - {}^-\tfrac{3}{4}x^2y$

9. $\quad \tfrac{1}{3}xy^2 - \tfrac{5}{8}xy^2 - {}^-\tfrac{1}{2}xy^2$

10. $\quad -\tfrac{3}{8}r^2s - \tfrac{2}{3}r^2s - -\tfrac{3}{4}r^2s - \tfrac{1}{2}r^2s$

11. $\quad x - 11 - -2x - -4 - 3$

12. $\quad 0.5a - 8 - {}^-4 - {}^-2a$

13. $\quad 9x - 3 - {}^-11x - {}^-4 - {}^-1$

14. $\quad {}^-\tfrac{1}{2}xy^2 - \tfrac{1}{4}x^2y - \tfrac{1}{3}xy^2 - \tfrac{2}{3}x^2y$

54

15. $-\frac{1}{4}a^2b - \frac{3}{4}ab^2 - {}^-\frac{2}{3}a^2b - \frac{1}{3}ab^2$

16. $x^2 - 2x - 5 - 3x - 1$

17. $2x^2 - x - {}^-x^2 - 2x - 2$

18. $.3y^2 - .9y - {}^-.2y^2 - {}^-.3y - .8$

19. $\frac{2}{3}x^2 - \frac{1}{3}x - {}^-\frac{7}{8}x^2 - 2x - 2$

20. $y^2 - {}^-\frac{1}{8}y - {}^-\frac{1}{4}y^2 - {}^-\frac{1}{2}y - {}^-3$

In other problems in the next section you will see chains of addition and subtraction links. This will transform your hybrid addition-subtraction chain into an addition chain.

Subtracting Binomials and Polynomials

This may be the most confusing section in all of algebra, so read carefully. In this section, polynomials have been written with minimal signs such that subtraction of positive terms will appear in both polynomials. As long as you remember that a minus sign is attached to the term that follows it, you will not err.
1.   **Keep** the signs on the first polynomial;
2.   Change the minus operation to a **plus** operation;
3.   **Opposite-opposite-opposite** every term in *the subtrahend* polynomial by **changing the symbol in front of it to the opposite**.
4.   Go back to the first polynomial and change only subtraction of positives monomials to addition of negative monomials. Never change the sign of the first monomial of the first polynomial.

Like terms have still been placed in columns. Once the problem has become addition, use Addition Rule # 1 or Addition Rule # 2. Never rewrite the order of minuend and subtrahend. Rather, apply Addition Rule # 2 mentally to like terms. To finalize your solution, convert addition of negative terms to subtraction of positive terms. This last idea bears repeating: **convert addition of negative terms to subtraction of positive terms.** This minimizes signs and operation symbols. By now the student is ready to accept the idea that an addition chain may contain subtraction because subtraction is addition of the opposite.

Ex.

$$6x^2 - 5 \quad = \qquad 6x^2 + -5$$
$$\underline{-\quad 3x^2 - 3} \qquad \underline{+ \quad -3x^2 + \quad 3}$$
$$\qquad\qquad\qquad 3x^2 + -2 \quad = \quad 3x^2 - 2$$

The procedure is logical because you are subtracting three x-squared and subtracting negative three from the first polynomial; these become adding negative three x-squared and adding positive three, respectively. For a review, adding binomials and polynomials begins on page 25.

1. $\quad\quad 3x^2 + 5$
   $\underline{-\quad 6x^2 + 3}$

2. $\quad\quad 4x^2 + \ 1$
   $\underline{-\quad 3x^2 - \ 4}$

3. $\quad\quad -5x^2 + 3x - \ 8 \quad$ (Hint: convert this to adding -8)
   $\underline{-\quad\quad 2x^2 - \ x + \ 9}$

4. $\quad\quad 12x^2 - 4x + 4$
   $\underline{-\quad -10x^2 - 5x + 2}$

5. $\quad\quad -7x^2 - 9x - 1$
   $\underline{-\quad -3x^2 + 5x - 1}$

6. $\quad\quad 10x^2 + 11x + 1$
   $\underline{-\quad -5x^2 - \ 6x + \ 1}$

7. $\quad\quad 2x^2 - \ 7x + \ 5$
   $\underline{-\quad -6x^2 - \ 4x - \ 2}$

8. $\quad\quad 4x^2 + 10xy + \ 9y^2$
   $\underline{-\quad 16x^2 + 32xy + 16y^2}$

9. $\quad\quad x^3 + \ 3x^2y + \ 3xy^2 + \quad y^3$
   $\underline{-\quad x^3 - \ 9x^2y - \ 9xy^2 - \ 27y^3}$

10. $\quad\quad x^4 + \ 4x^3y + \ 6x^2y^2 + \quad 4xy^3 + \quad y^4$
    $\underline{-\quad x^4 - \ 8x^3y - 24x^2y^2 - \ 32xy^3 + 16y^4}$

Subtracting Radicals

As with subtracting monomials, the subtraction chains in the problems below will be transformed into addition chains using **keep-plus-opposite-plus-opposite-plus-opposite** for as many links as there are in the chain. In this section, signs and symbols have *not* been minimized as in the previous section. Subtracting a positive means adding a negative, so turn the minus sign into plus and put a negative sign in front of a subtrahend that has no sign in front. Subtracting a negative means adding a positive, so make the double- minus a double-plus. Problems 1 through 6 contain all like terms. Problem 7 has two different kinds of like terms. Problems 8 through 10 have three different kinds of like terms. Problems 11 through 20 require converting into simplest radical form before like terms can be combined. Finalize your solutions by converting addition of negative terms in the chain to subtraction of positive terms. Adding radicals begins on page 27.

1. $2\sqrt{2}--3\sqrt{2}-5\sqrt{2}-7\sqrt{2}$

2. $5\sqrt{3}--4\sqrt{3}--2\sqrt{3}-7\sqrt{3}$

3. $\sqrt{5}--4\sqrt{5}--8\sqrt{5}-12\sqrt{5}$

4. $x\sqrt{3x}-2x\sqrt{3x}--4x\sqrt{3x}-9x\sqrt{3x}$

5. $4x^2\sqrt{x}+3x^2\sqrt{x}+-6x^2\sqrt{x}+-13x^2\sqrt{x}$

6. $5\sqrt{3}+2\sqrt{5}+-3\sqrt{3}+-7\sqrt{5}$

7. $-2\sqrt{2x}+3x\sqrt{2x}+-3\sqrt{2x}+-5x\sqrt{2x}$

8. $-6x\sqrt{2}+x\sqrt{3}+-4x\sqrt{2}+-3x\sqrt{3}+3\sqrt{2}$

9. $-9x\sqrt{3x}+3\sqrt{3x}+-5x\sqrt{3x}+4\sqrt{3x}+-12x\sqrt{3x}+7\sqrt{2x}$

10. $8x^2\sqrt{5x}+-3x\sqrt{2x}+-4x^2\sqrt{5x}+-23x\sqrt{5x}+x^2\sqrt{5x}+-4x^2\sqrt{5x}$

11. $\sqrt{8}+\sqrt{50}$

12. $\sqrt{18}+\sqrt{72}$

13. $\sqrt{27}+\sqrt{48}+\sqrt{147}+\sqrt{192}$

14. $\sqrt{75}+\sqrt{192}+\sqrt{432}+\sqrt{363}$

15. $\sqrt{80}+\sqrt{180}+\sqrt{245}$

16. $\sqrt{150}+\sqrt{384}$

17. $2x\sqrt{288}+\sqrt{338}+-3x\sqrt{72}$

18. $\sqrt{175} + -\sqrt{150} + -\sqrt{63}$

19. $-\sqrt{24} + x\sqrt{96} + \sqrt{20}$

20. $-\sqrt{24} + \sqrt{96} + \sqrt{20}$

Subtracting Algebraic Fractions

First apply **keep-plus-opposite**.   Once the problem has been converted to addition, **add** numerators using Addition Rule # 1 or Addition Rule # 2 and **keep** the common denominator. Fractions may need to be converted to like denominators as with addition of algebraic fractions, which begins on page 33.

Fraction	x One	=	Fraction of Same Value

$\dfrac{-3}{4} = \dfrac{-3}{4}$ x $\dfrac{3x}{3x}$ (multiply numerators) = $\dfrac{-9x}{12x}$ (multiply denominators)

$-$ $+$

$\dfrac{5}{6x} = \dfrac{-5}{6x}$ x $\dfrac{2}{2}$ (multiply numerators) = $\dfrac{-10}{12x}$ (multiply denominators)

$+$

$\dfrac{9x - 10}{12x}$

Add numerators · **Keep** common denominator

1. $\dfrac{3x}{4}$

$-$ $\dfrac{2x}{5}$

2. $\dfrac{-3x}{x+4}$

$-$ $\dfrac{2x}{x+4}$

3. $\dfrac{1}{4x}$

$-$ $\dfrac{-1}{6x}$

58

Fractions	x One	=	Fractions of Same Value

4. $\dfrac{-1}{5x}$

$-$

$\dfrac{-1}{8x}$

5. $\dfrac{-x}{x^2 - 5x + 6}$

$-$

$\dfrac{-2x}{x^2 - 5x + 6}$

6. $\dfrac{-3}{5x}$

$-$

$\dfrac{-5}{4x^2}$

7. $\dfrac{-2}{3x}$

$-$

$\dfrac{-7}{4x}$

8. $\dfrac{8}{3x}$

$-$

$\dfrac{-7}{5x^2}$

9. $\dfrac{-4}{9}$

$-$

$\dfrac{-5}{6x^2}$

10.
$$\frac{4}{3x^2}$$

$-$

$$\frac{2}{x}$$

When subtracting fractions requires using the distributive property in the numerator, the sign change to the second fraction is distributed to every term in the numerator. After applying **keep-plus-opposite** to every term in the numerator, the procedure is the same as for adding algebraic fractions.

Fractions		x One	=	Fractions of Same Value

$$\frac{-6}{x-3} = \frac{-6}{x-3} \bullet \frac{(x+2)}{(x+2)} = \frac{-6(x+2)}{(x-3)(x+2)}$$

$-$ $+$ $+$

$$\frac{-9}{x+2} \quad \frac{+9}{x+2} \bullet \frac{(x-3)}{(x-3)} = \frac{9(x-3)}{(x-3)(x+2)} \quad ,$$

$$\frac{-6(x+2)}{(x-3)(x+2)} + \frac{9(x-3)}{(x-3)(x+2)} = \frac{-6x+(-6 \bullet 2)+9x+(9 \bullet -3)}{(x-3)(x+2)}$$

$$= \frac{-6x+-12+9x-27}{(x-3)(x+2)} = \frac{3x-39}{(x-3)(x+2)}$$

Fractions		x One	=	Fractions of Same Value

11.
$$\frac{-5}{x}$$

$-$

$$\frac{-2}{2x-3}$$

12.
$$\frac{7}{x+8}$$

$-$

$$\frac{-5}{x-6}$$

60

Fractions	x One	=	Fractions of Same Value

13.
$$\frac{6}{x + 5}$$

$-$

$$\frac{-8}{x}$$

14.
$$\frac{5}{x + 11}$$

$-$

$$\frac{-2}{x}$$

15.
$$\frac{-3}{x - 7}$$

$-$

$$\frac{8}{x - 6}$$

16.
$$\frac{-4}{x - 2}$$

$-$

$$\frac{8}{x - 5}$$

61

Fractions	x One	=	Fractions of Same Value

17. $\dfrac{-3}{x+1}$

$-$

$\dfrac{8}{x-4}$

18. $\dfrac{-2}{x-7}$

$-$

$\dfrac{8}{x-6}$

19. $\dfrac{-5}{x-2}$

$-$

$\dfrac{8}{x-5}$

20. $\dfrac{-4}{x^2+1}$

$-$

$\dfrac{3x}{x-4}$

MULTIPLICATION

Multiplication is repeated addition. Six times five is simply

$$6 \cdot 5 = 5 + 5 + 5 + 5 + 5 + 5$$

Notice that this is an addition chain of six appearances of five, or five, six times. The answer is the same if we multiply five times six.

$$5 \cdot 6 = 6 + 6 + 6 + 6 + 6$$

Hence, multiplication is commutative.

The Vocabulary of Multiplication

factor · factor = product

factor - a term being multiplied to another.
product - the result of multiplication.

The Commutative Property of Multiplication

ab = ba   The order of factors does not affect the product.

Unlike addition and subtraction, there are many ways to express multiplication symbolically. To review, the only symbol for addition is "+." The only symbol for subtraction is " – ." The following, by contrast, are all representations of multiplication:

5 x 6	the familiar "x"
5 · 6	the dot (in type, an asterisk, *, may be used in its place)
5 (6)	one term outside of parentheses with no operation symbol in between
(5)(6)	two sets of parentheses with no symbol in between the sets
5x	means "five times the variable x."

When a number is being multiplied to a variable, the number is the first factor and is given the name **coefficient**. When a coefficient is placed before the variable **with no symbol in between**, the operation implied between the two is multiplication, and *only* multiplication. Therefore

$$5x$$

means 5 times the variable x or

$$x + x + x + x + x$$

It does not represent the incomplete expression "5 times (fill in the blank)." It is confusing that x is both a symbol for multiplication and also a variable. Many students represent 5x as 5 x x, which ambiguously could represent the multiplication chain of $5x^2$. Hence the convention of algebra asks students to refrain from using x as the symbol for multiplication and replace it with the dot, $\cdot$, with parentheses, or to omit a symbol completely and express a multiplication chain as a monomial.

Exponentiation is repeated multiplication in the same way that multiplication is repeated addition. So

$$x^5$$

means x appears on five occasions in the multiplication chain or

$$x \bullet x \bullet x \bullet x \bullet x$$

Consider the following table of values contrasting repeated addition with repeated multiplication.

Integer x	Multiplication $x + x = 2x$	Exponentiation $x \cdot x = x^2$
- 3	$-3 + -3 = 2(-3) = -6$	$(-3)(-3) = (-3)^2 = 9$
- 2	$-2 + -2 = 2(-2) = -4$	$(-2)(-2) = (-2)^2 = 4$
- 1	$-1 + -1 = 2(-1) = -2$	$(-1)(-1) = (-1)^2 = 1$
0	$0 + 0 = 2(0) = 0$	$(0)(0) = 0^2 = 0$
1	$1 + 1 = 2(1) = 2$	$(1)(1) = 1^2 = 1$
2	$2 + 2 = 2(2) = 4$	$(2)(2) = 2^2 = 4$
3	$3 + 3 = 2(3) = 6$	$(3)(3) = 3^2 = 9$

As stated in the subchapter on adding radicals, the fact that four is both the sum and product of two repeated causes students great confusion and persistent errors in squaring and taking square roots of integers; it is ironic, considering that two together with zero are the only integers to exhibit this phenomenon. Students who confuse ordinary multiplication with exponentiation must now commit to permanent, long-term memory the distinction between the first, which is repeated addition, and the second, which is repeated multiplication.

Multiplication and addition, unlike subtraction and division, are both commutative and associative.

The Commutative Property

$a + b = b + a$ The order of addends does not affect the sum.

$ab = ba$ The order of factors does not affect the product.

The Associative Property of Addition

$a + (b + c) = (a + b) + c$ When forming an addition chain, like terms from one or more chains may be regrouped in order to be combined into one term.

Ex. $x + (x + 2) = (x + x) + 2 = 2x + 2$

The Associative Property of Multiplication

$a(bc) = (ab)c$ When forming a multiplication chain, terms from one or more chains may be regrouped in order to express the chain in simplest form.

Ex. $2(6.022 \times 10^{23}) = (2 \times 6.022) \times 10^{23} = 12.044 \times 10^{23}$
$= 1.2044 \times 10^{24}$

Multiplication differs from addition in that, whereas addition is the taking of inventory and the consolidation of only like terms into a single monomial, *multiplication usually changes the nature of the terms being multiplied.* When length is added to or subtracted from length, the new term remains length. But when length is *multiplied* to length, the new term becomes area, a two-dimensional term, and length multiplied to area becomes volume, a three-dimensional term.   Likewise, when the rate of an object's speed is multiplied by the time it travels, the product is the distance traveled by the object, a term with a completely different nature. Because multiplication is not dependent upon the nature of the terms being multiplied, unlike terms can ALWAYS be linked together in a multiplication chain to produce a new monomial with a different nature or dimension; then, unlike monomials are linked together in an addition chain to produce a polynomial.

multiplication chain + multiplication chain − multiplication chain + multiplication chain

↑

addition chain polynomial

The rules for multiplying all things mathematical are easy to commit to permanent, long-term memory.

Multiplication Rule # 1

(1)   When **two** factors, or an **even number of factors**, have the **SAME signs**, their product is **POSITIVE**.

Pairs of Positive Factors Have the
Same Positive Sign and are Multiplied
To Produce Positive Products

$$3 \cdot 5 = 15$$
$$1 \cdot 9 = 9$$
$$17 \cdot 24 = 408$$
$$23 \cdot 45 = 1,035$$
$$.23 \cdot .45 = 0.1035$$

$$\frac{2}{7} \cdot \frac{4}{7} = \frac{8}{49}$$

$$2x \cdot 4x = 8x^2$$

$$2\sqrt{2} \cdot 4\sqrt{2} = 8\sqrt{4} = 16$$

66

Pairs of Negative Factors Have the
Same Negative Sign and are Multiplied
To Produce Positive Products

$$^-3 \cdot {}^-5 = 15$$
$$^-1 \cdot {}^-9 = 9$$
$$-17 \cdot -24 = 408$$
$$-23 \cdot -45 = 1{,}035$$
$$-.23 \cdot -.45 = 0.1035$$

$$\frac{-2}{7} \cdot \frac{-4}{7} = \frac{8}{49}$$

$$^-2x \cdot {}^-4x = 8x^2$$
$$^-2\sqrt{2} \cdot {}^-4\sqrt{2} = 8\sqrt{4} = 16$$

Multiplication Rule # 1 Explained

Ex. 1. $-3 \cdot -5$ "I owed five friends three dollars each but
they all cancelled my debt. I have fifteen dollars."

Each debt of three dollars has been negated:

$$-(-3) + -(-3) + -(-3) + -(-3) + -(-3)$$
$$= 3 + 3 + 3 + 3 + 3 = 15$$

Ex. 2: $-2 \cdot -3 \cdot -2 \cdot -5 = 60$

By the associative property of multiplication,

$$(^-2 \cdot {}^-3) \cdot (^-2 \cdot {}^-5) = 6 \cdot 10 = 60$$

Since two negative factors produce a positive product, any pair or even number of negative factors in a multiplication chain of positive and negative factors together will always produce a positive product.

Multiplication Rule # 2

(2) When **two** factors have **DIFFERENT signs**, or an **odd number of factors are negative**, their product is **NEGATIVE**.

67

The second part of the rule follows from the first part.

Multiplication Rule # 2 Explained

Ex. 3. $^-3 \cdot 5 = {}^-15$ "I owe five friends three dollars each.
I owe fifteen dollars altogether."

Since multiplication is repeated addition,

$$-3 + -3 + -3 + -3 + -3 = -15$$

(from Addition Rule # 1). Also, since multiplication is commutative, placing the second factor first will produce the same product.

Ex. 4. $1 \cdot {}^-9 = {}^-9$

The multiplicative identity property is entirely consistent with all operations rules and, more specifically, with the second multiplication rule. If negative 9 appears once, it equals itself.

Ex. 5. $-5 \cdot -3 \cdot -15 = -225$

By the associative property of multiplication, negative five times negative three equals positive fifteen. This term times negative fifteen equals negative two hundred twenty-five.

A table may be used to illustrate the multiplication rules.

+	x	+	=	+
+	x	−	=	−
−	x	+	=	−
−	x	−	=	+

In any multiplication chain, the rules of the table are applied to the sign of the first two factors multiplied; the product then is multiplied to the next factor following the same rules. This continues until the final product is obtained. The sign on that product will be positive if an even number of negative factors are in the chain, and negative if an odd number of negative factors are in the chain.

68

Multiplying Integers

1. $4 \cdot {}^-6 =$

2. ${}^-4 \cdot {}^-6 =$

3. $4 \cdot 6 =$

4. $-4 \cdot 6 =$

5. $3 \cdot -7 =$

6. $-3 \cdot 7 =$

7. $-3 \cdot -7 =$

8. $-8 \cdot -5 =$

9. $8 \cdot -5 =$

10. $-8 \cdot 5 =$

11. $2 \cdot -9 =$

12. $-2 \cdot 9 =$

13. $2 \cdot -9 =$

14. $-2 \cdot -9 =$

15. $-11 \cdot 4 =$

16. $11 \cdot -4 =$

17. $-11 \cdot -4 =$

18. $-10 \cdot -2 =$

Multiplying Numerical Fractions

You have already visited the procedure for multiplying fractions when you were adding fractions with unlike denominators. To multiply fractions, **multiply numerators** to form a product numerator and **multiply denominators** to form a product denominator. To obtain the correct sign on the product, use Multiplication Rule # 1 or Multiplication Rule # 2:

(1) When **two** factors, or any **even number of factors**, have the **SAME signs**, their product is **POSITIVE**.

(2) When **two** factors have **DIFFERENT signs**, or any **odd number of factors are negative**, their product is **NEGATIVE**.

1. $\dfrac{-5}{3} \bullet \dfrac{-2}{3}$

2. $\dfrac{7}{8} \bullet \dfrac{-5}{6}$

3. $\dfrac{8}{5} \bullet \dfrac{-8}{3}$

4. $\dfrac{3}{5} \bullet \dfrac{6}{23}$

5. $\dfrac{-3}{8} \bullet \dfrac{5}{7}$

6. $\dfrac{-1}{4} \bullet \dfrac{2}{27}$

7. $\dfrac{-2}{9} \bullet \dfrac{-8}{21}$

8. $\dfrac{-4}{3} \bullet 5$ [Hint: By the Multiplicative Identity Property, the invisible denominator of any integer, variable or other term is one, because anything divided by one equals itself.]

Often, product fractions result which are not in simplest form. Consider

$$\dfrac{-6}{5} \bullet \dfrac{-8}{3}$$

By multiplying numerators to form a new product numerator, multiplying denominators to form a new product denominator, and using the multiplicative identity property in its third usage to factor and reduce the fraction to simplest form, we obtain

$$\dfrac{-6}{5} \bullet \dfrac{-8}{3} = \dfrac{48}{15} = \dfrac{16}{5} \bullet \dfrac{3}{3} = \dfrac{16}{5}$$

But there is a better way: instead of multiplying before and reducing after, reduce before and multiply after:

$$\dfrac{-\overset{-2}{\cancel{6}}}{5} \bullet \dfrac{-8}{\underset{1}{\cancel{3}}} = \dfrac{16}{5}$$

This works because, by the multiplicative identity property, anything times one equals itself. When a fractional equivalent of one is factored from the fraction, the fraction is reduced. Factoring out the largest possible fractional equivalent of one will result in a fraction reduced to simplest form. The shortcut for this procedure is to divide any factor in the numerator and any factor in the denominator by the greatest common factor between them.

70

Often, these common factors will be found by examining a diagonal path between a numerator factor and a denominator factor, sometimes because the problem turns out that way, and sometimes because the writer of the problem is testing the student on his understanding of the commutative and identity properties of multiplication.

9. $\dfrac{9}{2} \bullet \dfrac{-8}{3}$

10. $\dfrac{3}{11} \bullet \dfrac{6}{33}$

11. $\dfrac{-3}{8} \bullet \dfrac{8}{30}$

12. $\dfrac{-1}{4} \bullet \dfrac{4}{27}$

13. $\dfrac{-16}{15} \bullet \dfrac{-5}{6}$

Multiplying Decimals

To multiply decimals, temporarily ignore the decimals and place the last digits of factors above each other; the factor which has more digits should be placed above first. Multiply using the two rules of multiplication.
(1)   When **two** factors, or any **even number of factors**, have the **SAME signs**, their product is **POSITIVE**.
(2)   When **two** factors have **DIFFERENT signs**, or any **odd number of factors are negative**, their product is **NEGATIVE**.

The number of places after the product's decimal will equal the sum of the decimal places of the two factors.

Ex.
```
          4,398.0827     ← four decimal places
      x      - 32.98     ← two decimal places
         351846616
        3958274430
        8796165400
      + 131942481000
   - 145,048.767446      ← six decimal places
```
71

Notice that addition of rows of products is essential in order to obtain the final product. This concept is lost on students who only used calculators to perform multiplication. The indispensability of addition as part of multiplication will be seen repeatedly as the chapter progresses. Try the following problems without a calculator and check with it. Notice that a zero digit in the second factor eliminates multiplication for that entire row but requires an additional zero as a place holder in the next row.

1. -83745.87
    $\underline{\text{x} \qquad 43.409}$

2. -75403.97345
    $\underline{\text{x} \qquad -47.92}$

3. -3429.087
    $\underline{\text{x} \qquad 2.01}$

Multiplying Numbers Expressed in Scientific Notation

To multiply numbers expressed in scientific notation:

MULTIPLY the coefficients – **KEEP** the base ten – **ADD** the exponents
using Multiplication Rule   using Addition
    # 1 or # 2   Rule # 1 or # 2

Consider the following example

$(3.29453 \times 10^4)(5.2356 \times 10^6)$

By the associative property of multiplication, factors in the multiplication chain are regrouped to form a single monomial.

$= (3.29453)(5.2356)(10^4)(10^6)$

$= (17.248841268)(10 \times 10 \times 10 \times 10)(10 \times 10 \times 10 \times 10 \times 10 \times 10)$

Simplifying, we obtain

$= 17.248841268 \times 10^{10} = 1.7248841268 \times 10^{11}$

Recall that the decimal must be adjusted one place to the left and the exponent increased by one in order to conform the coefficient to the convention of being expressed as a decimal between the numbers one and ten. The shortcut is to multiply coefficients using Multiplication Rule # 1, to keep the base ten, and to add exponents using Addition Rule # 1.

The following problem is similar except that it requires the application of Multiplication Rule # 2. The shortcut will be taken.

$(4.8345 \times 10^6)(-9.21023 \times 10^3) = -44.526856935 \times 10^9$
$= -4.4526856935 \times 10^{10}$

Consider the following multiplication problem.

$(4.20398 \times 10^{-7})(7.3567853038 \times 10^2)$

The term 10^{-7} equals $\dfrac{1}{10^7}$ because a power with a negative exponent
equals the reciprocal of that power with the sign of the exponent changed to
positive.   This in turn means one-tenth appearing seven times in a
multiplication chain, or (.1)(.1)(.1)(.1)(.1)(.1)(.1). The decimal equivalent of
this is .0000001 or one ten-millionth.   Note that the negative exponent
indicates the number of places to the *left* of the number one (or of the
coefficient) that the decimal would have to be moved in order to convert the
term from scientific notation to standard (decimal) form. A comprehensive
visual comparison of positive, zero and negative exponents can be found in
Appendix B; moreover, zero and negative exponents are discussed in detail
in the subchapters on dividing numbers in scientific notation and dividing
algebraic monomials. For now, the student need only memorize that any
base to the zero power equals 1, and any base to a negative power equals the
reciprocal of that power with the sign of the exponent changed to positive.

By the associative property of multiplication, the terms in the multiplication
chain are regrouped:

$(4.20398 \times 7.3567853038) \times (10^{-7} \times 10^2)$

$= 30.927778281469124 \times \dfrac{1}{10} \times \dfrac{1}{10} \times \dfrac{1}{10} \times \dfrac{1}{10} \times \dfrac{1}{10} \times \dfrac{1}{10} \times \dfrac{1}{10} \times 10 \times 10$

By the multiplicative inverse property, ten multiplied by its reciprocal equals
one, the two factors of ten and two factors of one-tenth reduce to one; by the
first usage of the multiplicative identity property, one vanishes from the
multiplication chain.

$= 30.927778281469124 \times \dfrac{1}{10} \times \dfrac{1}{10} \times \dfrac{1}{10} \times \dfrac{1}{10} \times \dfrac{1}{10}$

$= 30.927778281469124 \times 10^{-5} = 3.0927778281469124 \times 10^{-4}$

Of course, the shortcut is the same as before, except that now Addition Rule # 2 must be applied to the exponents. So,

$$(4.20398 \times 10^{-7})(7.3567853038 \times 10^2) = 30.927778281469124 \times 10^{-5}$$

Coefficient is greater than ten; coefficient and exponent are adjusted to express the coefficient as a decimal between one and ten.

$$= 3.0927778281469124 \times 10^{-4}$$

Notice that the exponent was in fact increased by one, as is always the case whenever the decimal is moved to the left one place value. Try the following problems, using a calculator only to multiply coefficients; mentally add or subtract the exponents as in the illustrated example. Then adjust the decimal, if necessary, to follow the first digit of the coefficient. Adjust the exponent by adding positive one for each time you move the decimal in the coefficient left, and subtracting one for each time you move the decimal in the coefficient right. The various symbols used to express multiplication have been used here in order to help you recognize multiplication chains and distinguish them from addition/subtraction chains.

1. $(-1.3682253 \times 10^6)(-7.946783 \times 10^4)$

2. $-1.3682253 \times 10^6 (-7.946783 \times 10^{-8})$

3. $-4.4801438 \times 10^{-5} \bullet -1.09253874 \times 10^{-3}$

4. $-4.4801438 \times 10^2 x -2.095698712 \times 10^7$

5. $4.48014 \times 10^5 * -4.938053 \times 10^3$

6. $4.4801 \times 10^0 x -9.835 \times 10^9$

7. $(-3.2948 \times 10^{-5})(-9.5847 \times 10^3)$

8. $-9.765 \times 10^{-5} \bullet -1.57 \times 10^{12}$

9. $(4.4801438 \times 10^5)^2$
   [Note: rewrite this as the term in parentheses multiplied by itself.]

10. $(-8.465019325 \times 10^{-3})^2$

Multiplying Monomials

The Anatomy of a Monomial

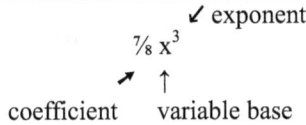

⟋ exponent

$\frac{7}{8} x^3$

⟋ ↑

coefficient   variable base

coefficient- a numerical factor always placed first in a multiplication chain.
power - a term with a base and an exponent on the base.
base - the object of repeated multiplication to itself; the repeating factor. When the base is a variable, the multiplication chain is expressed as a single term called a monomial.
variable - a letter used to represent an unknown or changing quantity. Its value is determined by its placement into an equation.
exponent- indicates the number of times the base appears in the multiplication chain.

interpreted as $\frac{7}{8} \cdot x \cdot x \cdot x$

This monomial represents four factors in a multiplication chain.

A monomial is an algebraic expression which may contain the operations of multiplication, division or exponentiation. A monomial never contains the operation of addition or subtraction. Examples of monomials follow:

$$7 \qquad -9 \qquad x \qquad 3x^2 \qquad \frac{4x^3}{3} \qquad {}^-y \qquad 5y^3$$

To multiply monomials:

MULTIPLY coefficients - **KEEP** the base - **ADD** the exponents

$$-3x \cdot 4x^2 = -3 \cdot x \cdot 4 \cdot x \cdot x = -3 \cdot 4 \cdot x \cdot x \cdot x = -12x^3$$
or
$$= -3 \cdot 4 \cdot x^{1+2} = -12x^3 \qquad \text{(Multiplication Rule \# 2/Addition Rule \# 1)}$$

By the associative property of multiplication, factors in the multiplication chain are regrouped to form a single monomial. Notice that the first monomial has an invisible exponent of one; this is added to the exponent of two to make an exponent of three. Do NOT confuse the process with combining like terms; this phrase is reserved exclusively for consolidation by means of addition or subtraction. Multiplication, by contrast, always permits consolidation of unlike terms into a single term called a monomial.

As in the first example, the two rules of multiplication must be followed when multiplying coefficients: if two coefficients have the same sign, the product is positive and if two coefficients have different signs the product is negative. When adding exponents, if the signs are the same, add using Addition Rule # 1 and if the signs are different, subtract using Addition Rule # 2.

$$-3x \cdot 4x^{-2} = -12 x^{1+-2} = -12x^{-1} = \frac{-12}{x} \quad \text{(Multiplication Rule \# 1}$$
$$\text{Addition Rule \# 2)}$$

An all-too-common error is

$$-3x^2 \cdot 4x^5 = -12x^{10}$$

Notice that the student has multiplied coefficients and also has multiplied exponents, collapsing the procedure for exponentiation of monomials discussed below with multiplication of monomials. The error is based on the belief that multiplication is an island unto itself, a misconception formed from relying on calculators to perform multiplication in elementary school. When a calculator is used to multiply, the only operation symbol pressed by the student is the multiplication key; when multiplication of multi-digit numbers is performed manually by the student, the necessity of adding rows of products to obtain the final product subconsciously becomes ingrained. Students who perform enough repetitive practice of multiplication without a calculator obtain the understanding of addition-as-part-of-multiplication; then, the concept easily is carried forward to multiplying monomials and other algebraic terms.

Thus, when approaching the problem

$$-7x^{-3} \cdot 4x^0$$

The solution is

$$-7x^{-3} \cdot 4x^0 = -28x^{-3} = \frac{-28}{x^3}$$

and not $-28 x^0$

76

When x is multiplied to $-x$, the result is $-x^2$.

$$x \bullet \; -1x = -x^2$$

1. $-2x \cdot 4x^5$

2. $-x^{-1} \cdot 7x^2$

3. $8x^0 \cdot -5x^{-2}$

4. $9x^{-4} \cdot -10x^{-2}$

5. $-11x^{-1} \cdot -3x^{-2}$

Sometimes monomials have more than one kind of base. The pattern **multiply-keep-add-keep-add-keep-add** will be used for finding such products. Remember to use Multiplication Rule # 1 or Multiplication Rule # 2 on the coefficients and Addition Rule # 1 or Addition Rule # 2 on the exponents of like bases.

$$2x^2y^2z \bullet x^2y \bullet y^3z^2 = 2 \bullet 1 \bullet 1 \bullet x^{2+2+0}y^{2+1+3}z^{1+0+2} = 2x^4y^6z^3$$

When any monomial contains parentheses around it and an exponent attached thereto, the entire monomial is to be multiplied by itself as indicated by the exponent:

$$(x^2)^5 = (x^2)(x^2)(x^2)(x^2)(x^2) = x^{2 \cdot 5} = x^{10}$$

$$(4x^2y^3)^2 = (4x^2y^3)(4x^2y^3) = 4 \bullet 4 \bullet x^2 \bullet x^2 \bullet y^3 \bullet y^3 = 4^{1 \cdot 2}x^{2 \cdot 2}y^{3 \cdot 2} = 16x^4y^6$$

The shortcut is to distribute the exponent to every coefficient and base in the multiplication chain inside the parentheses, **multiplying** the exponent to any existing exponents on such terms. This does *not* follow the multiply-keep-add pattern because *multiplying exponents* results from *exponentiation* of monomials whereas adding exponents results from multiplying monomials. Rather, the pattern here is to **exponentiate** the coefficient, **keep** the like base and **multiply** the exponent on the base to the exponent outside parentheses.

$$(4x^2y^3)^2 = 4^2 \, x^{2 \bullet 2}y^{2 \bullet 3} = 16x^4y^6$$

Bear in mind that in order to apply this rule, the chain inside parentheses must be strictly one of multiplication.

At this point it is important to illustrate how a monomial can be a multiplication chain which includes parenthetic binomials or longer addition chain polynomials. Consider

$$x^2(x + 3)(x^2 - 4)(3x - 5)$$

$$\uparrow \quad \uparrow \quad \uparrow$$

Multiplication links

Until now, we have seen addition chains comprised of unlike monomial multiplication chains. The above example indicates the reverse idea: a multiplication chain may contain links comprised of binomials and longer addition chain polynomials. When the *outer chain* is one of multiplication, the term is still a monomial. Rather than become intimidated by the seeming complexity of such an expression, the student should simply remain mindful of the *absence of any plus or minus signs linking sets of parentheses* in order to identify these multiplication chains. The rules for multiplication apply to the exponents, visible or invisible, attached to the parentheses.

$$x^5y^{-3}z(y + z) \bullet - 13x^2y^{-1}z\,(y + z)\bullet 3x^3y^2\,(x + y)$$

$$= -39\,x^{5+2+3}y^{-3+-1+2}z^{1+1}(y + z)^{1+1}(x + y)$$
$$= -39\,x^{10}y^{-2}z^2(y + z)^2(x + y)$$

$$= \frac{-39\,x^{10}\,z^2(x + y)(y + z)^2}{y^2}$$

6. $(-3x^2)^3$

7. $7x^3y^2z \bullet - 9x^2y \bullet 8y^{-3}z^2$

8. $-5a^{-2}b^2c \bullet b^2c \bullet 2a^3c^2$

9. $x^{-3}y^{-1}z \bullet 6x^{-5}y^{-3} \bullet - 8y^4z^2$

10. $p^2q^{-4}r \bullet p^2q \bullet q^5r^{-2}$

11. $(5x^3y^4z^5)^2(x + y + z)^2$

12. $h^5j^{-3}k \bullet - 7h^2j^{-1}k(j + k) \bullet 11j^3k^2\,(j + k)$

13. $(-6h^5j^{-3}k)^{-3}(h - j - k)\cdot - 4h^{11}j^{12}k(h - j - k)$

78

Multiplying a Monomial by a Polynomial

The Distributive Property of Multiplication over Addition
a (b + c) = ab + ac   a factor being multiplied to an addition chain is multiplied to each and every term, like or unlike, in the chain.
From this it follows
a(b + c − d) = ab + ac − ad

Consider the following basic multiplication problem

```
      51
x      4
     204
```

This is essentially

```
(50 + 1)
x      4
       4
+    200
     204
```

Notice that addition of rows of products is required to obtain the final product. It should come as no surprise that addition is a necessary and indispensable element of multiplication, given that multiplication is repeated addition.

Instead of multiplying vertically, we can multiply horizontally, placing four in front of the binomial (50 + 1).

4 (50 + 1)

Using the distributive property and following the order of operations whereby multiplication is honored before addition, this becomes

$4 \times 50 + 4 \times 1 = 200 + 4 = 204$

This problem illustrates the idea that the distributive property is simply ordinary multiplication, performed horizontally rather than vertically. The converse is also true, namely, all multiplication beyond single digit terms is multiplication over addition, which is the distributive property.

Now make a simple algebraic replacement: $(50 + 1)$ becomes $x + 1$

$$
\begin{array}{r}
x + 1 \\
\cdot \qquad 4 \\
\hline
+ \, 4 \\
+ \quad 4x \\
\hline
4x + 4
\end{array}
$$

Again we can multiply horizontally instead of vertically:

$4(x + 1) = 4x + 4 \bullet 1 = 4x + 4$

To reiterate, the distributive property is nothing more than horizontal multiplication involving the removal of parentheses in the process.

When distributing a negative term to a binomial with a subtraction symbol followed by an implied positive term, the second product becomes negative. Then, subtracting a negative becomes adding a positive.

$-3(x - 8) = -3x - (^-3)(8) = -3x - {}^-24 = -3x + 24$

1. $-6(x - 7)$

2. $5(x^2 - 4x + 3)$

3. $-4(x^2 - 2x + 1)$

4. $\sqrt{2} \, (4x^2 - 5x + 1)$

5. $5xy(x^2 + 2xy + y^2)$

6. $4z(x^4 - 8x^3y + 24x^2y^2 - 32xy^3 + 16y^4)$

Multiplying Binomials

To review, a binomial is nothing more than two unlike monomials connected by addition or subtraction in a two-link chain.

$$x + 7 \qquad\qquad x - 3 \qquad\qquad x^2 - 16$$

The example on the right is called a second degree or second order binomial because its monomial of highest power is two. Binomials at their most basic consist of a variable raised to the (invisible) first power, a constant and an addition or subtraction symbol in between the two.

Multiplying binomials requires use of the distributive property performed twice. Consider

$$
\begin{array}{r}
51 \\
\bullet \quad 14 \\
\hline
204 \\
+ \quad 510 \\
\hline
714
\end{array}
$$

Given that 51 equals 50 + 1 and 14 equals 10 + 4

$$
\begin{array}{r}
(50 + 1) \\
\bullet \ (10 + 4) \\
\hline
4 \\
200 \\
10 \\
+ \quad 500 \\
\hline
714
\end{array}
$$

one times four plus fifty times four plus
one times ten plus fifty times ten

Now make a substitution: replace fifty and ten with x:

$$
\begin{array}{r}
(x + 1) \\
\bullet \quad (x + 4) \\
\hline
4x + 4 \\
+ \ x^2 + 1x \\
\hline
x^2 + 5x + 4
\end{array}
$$

one times four plus x times four plus
one times x plus x times x

combine like terms

Multiplying two first order binomials results in a second order trinomial.

Performing the same operation horizontally rather than vertically, the outcome is the same.

$$(x + 1)(x + 4) = x \cdot x + x \cdot 4 + 1 \cdot x + 1 \cdot 4 \ = \ x^2 + 4x + x + 4$$

First   Outer   Inner   Last   combine like terms
products products products products

$$= x^2 + 5x + 4$$

The popular acronym FOIL – for First products, Outer products, Inner products, Last products – is used by students and teachers alike as both a noun ("multiply using FOIL") and a verb ("FOIL it out") to refer to multiplication of binomials. Note that when two linear binomials are multiplied, two of the products are linear monomials which should always be combined into a single linear monomial; it then is linked to the quadratic monomial and constant in an addition chain to form the unity of a quadratic trinomial.

So far we have seen two different methods of display; regardless of the method, the outcome is the same. Nevertheless, the following methods are alternatives and are simply a matter of student preference.

$$\overset{\text{F}\quad\text{O}}{(x + 1)(x + 4) = x^2 + 4x}$$
$$\underset{\text{I}\quad\text{L}}{}$$
$$\underline{\qquad\qquad + x + 4}$$
$$x^2 + 5x + 4$$

or

$$(x + 1)(x + 4) =$$

F	$x \cdot x = x^2$
O	$4 \cdot x = 4x$ ⎤ combine
I	$1 \cdot x = \ x$ ⎦ like terms
L	$1 \cdot 4 = \ 4$

$$= x^2 + 5x + 4$$

Lastly, we come to the matrix or mini-times table. Place the monomials with the front-attached operation symbol above or next to each box and fill in each product using **multiply-keep-add**. Then combine the two like terms along the diagonal using the addition rule **add-keep-keep**.

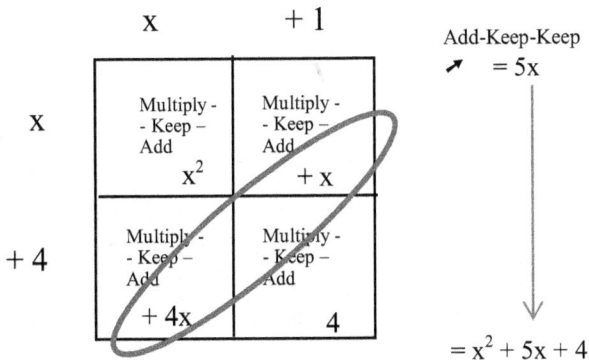

	X	+ 1
X	Multiply - - Keep – Add x^2	Multiply - - Keep – Add $+ x$
+ 4	Multiply - - Keep – Add $+ 4x$	Multiply - - Keep – Add 4

Add-Keep-Keep

↗ = 5x

$= x^2 + 5x + 4$

1.　　$(x - 9)(x + 6)$

2.　　$(x + 4)(x + 5)$

3.　　$(x - 6)(x - 6)$

(Note: this results in a *perfect square trinomial* because the binomials are equal; when two equal terms are multiplied together they form, from a geometric perspective, a perfect square.)

4.　　$(x - 8)(x + 4)$

5.　　$(2x + 7)(4x - 5)$

6.　　$(x + 4)(x - 4)$

(Note: this results in a *difference of perfect squares*, not to be confused with the perfect square trinomial, because the inner and outer products add up to zero, leaving a second degree binomial, each of whose monomials are a perfect square.)

83

Multiplying Polynomials

Multiplying polynomials requires use of the distributive property as many times as there are monomials in the shorter polynomial; each monomial in the first polynomial will be multiplied to each monomial in the second polynomial. Consider

$$(x^2 - 3x + 2)(x^2 - 5x + 6)$$

This is a trinomial being multiplied by another trinomial. The distributive property must be used three times. As always, apply **multiply-keep-add**, using Multiplication Rule # 1 or Multiplication Rule # 2 on the coefficients and Addition Rule # 1 or Addition Rule # 2 on the exponents:

$$
\begin{array}{r}
x^2 - 3x + 2 \\
\bullet\, x^2 - 5x + 6 \\
\hline
6x^2 - 18x + 12 \\
- 5x^3 + 15x^2 - 10x \\
x^4 - 3x^3 + 2x^2 \\
\hline
x^4 - 8x^3 + 23x^2 - 28x + 12
\end{array}
$$

Students may find using the matrix less prone to error. Use one column for each monomial in the first polynomial and one row for each monomial in the second polynomial. Combine like terms by adding along the diagonals.

	x^2	$- 3x$	$+ 2$
x^2	Multiply-Keep-Add x^4	Multiply-Keep-Add $- 3x^3$	Multiply-Keep-Add $2x^2$
$- 5x$	Multiply-Keep-Add $- 5x^3$	Multiply-Keep-Add $15x^2$	Multiply-Keep-Add $- 10x$
$+ 6$	Multiply-Keep-Add $6x^2$	Multiply-Keep-Add $- 18x$	Multiply-Keep-Add 12

$$= x^4 - 8x^3 + 23x^2 - 28x + 12$$

1. $(x^2 + 2)(x^2 - 4x + 4)$

84

2. $(4x^2 + 1)(-5x^2 + 3x + -8)$

3. $(x - 8)(-x^3 + 2x^2 - 3x - 8)$

4. $(x - 13)(-2x^3 - 3x^2 - 4x - 5)$

5. $(x + 1)(4x^3 + x^2 - x - 7)$

Multiplying Radicals

Consider

$$\sqrt{4 + 9} \text{ versus } \sqrt{4 \cdot 9}$$

Clearly, the second radical equals the square root of 36 which equals six. Likewise, the second radical can be broken up into the multiplication chain

$$\sqrt{4}\sqrt{9} = 2 \cdot 3 = 6$$

producing the same result, because the radical of a product equals the product of the individual radical factors. Hence, radicals of a multiplication chain can be converted into multiplication chains of radical factors of the original radicand. We saw this in the subchapters on adding and subtracting radicals. In contrast, radical four *plus* nine equals radical thirteen which is approximately equal to three-point-six-zero- five-five-five-one-two- seven-five which is *not* equal to two plus three which equals five.

$$\sqrt{4+9} = \sqrt{13} \approx 3.6055512755 \neq 2 + 3 = 5$$

The radical of a sum or difference is NOT equal to the sum or difference of the individual radical addends. It is necessary, rather, to think of a sum or difference under a radical as being a unity subjected to the operation of finding its root. The method of undoing the radical of an addition chain will be explored when we solve radical equations.

From the example above it follows that multiplication chains of factors can sometimes be lengthened and shortened to produce the radical of a perfect square. For example, the square root of three alone is irrational, as is the square root of seventy-five. But, multiplied together, they equal fifteen.

$$\sqrt{3}\sqrt{75} = \sqrt{3 \cdot 75} = \sqrt{225} = 15$$

Often, multiplying radicals with variables in the radicands will result in a power whose root can be taken.

$$\sqrt{3x}\sqrt{75x} = \sqrt{3 \cdot 75x^2} = \sqrt{225x^2} = 15x$$

Recall that $x \cdot x = x^2$. In fact, every even variable power is a perfect square; its root is the variable with the *exponent divided by two*. Do not collapse this procedure with finding the root of a numerical *base*, which is never accomplished with division by two.

So, $x^2 \cdot x^2 = x^4$ thus $\sqrt{x^4} = x^2$
and $x^3 \cdot x^3 = x^6$ thus $\sqrt{x^6} = x^3$
and $x^4 \cdot x^4 = x^8$ thus $\sqrt{x^8} = x^4$

When the variable in the radicand has an odd exponent greater than one, subtract one from the exponent and divide the remaining even exponent by two; the singular variable remains in the radicand and the root of the even power comes out of the radical.

86

Thus $\sqrt{x^7} = \sqrt{x^6 \cdot x} = x^3\sqrt{x}$

Try these problems.

1. $\sqrt{3x}\sqrt{48x}$

2. $\sqrt{2}\sqrt{128x}$

3. $\sqrt{2x}\sqrt{3x}\sqrt{54x}$

4. $\sqrt{2x}\sqrt{98x^2}$

5. $\sqrt{3}\sqrt{192}$

6. $\sqrt{2}\sqrt{2}\sqrt{144}$

7. $\sqrt{5}\sqrt{125}$

Often, radicals have coefficients. These multiplication chains usually can be simplified. Multiplying coefficients and multiplying radicands gives a radical which can be reduced to simplest form.

$$6\sqrt{7} \cdot 2\sqrt{14} = 12\sqrt{98} = 12\sqrt{49 \cdot 2} = 12 \cdot 7\sqrt{2} = 84\sqrt{2}$$

Again, recall that to simplify radicals, divide by trial and error the radicand by 2, 3, 5, 6, or 7. One of these numbers will be the remaining radicand; the other number will be the largest possible perfect square whose square root can then be taken and converted into an integral factor, placed in front of the radical and creating a multiplication chain. Try dividing by two first; if the other number is not a perfect square, try to divide the radicand by three, and so on, until a perfect square is obtained.

In the same way that perfect squares can be converted to integers and simple fractions, integers and simple fractions can be pushed into a radical to create a perfect square. Be aware, however, that this technique has limited utility and should be reserved for dividing radicals; it is often better to multiply coefficients, multiply radicands and then reduce to simplest form.

Fine: $2\sqrt{2} \bullet \sqrt{32} = \sqrt{4 \bullet 2}\sqrt{32} = \sqrt{8}\sqrt{32} = \sqrt{256} = 16$

Better: $2\sqrt{2} \bullet \sqrt{16}\sqrt{2} = 2 \bullet 4 \bullet \sqrt{2}\sqrt{2} = 8 \bullet 2 = 16$

8. $2\sqrt{24}\sqrt{6}$

9. $6\sqrt{5}\cdot2\sqrt{15}$

10. $2\sqrt{2}\sqrt{72}$

11. $\sqrt{8}\sqrt{18}$

12. $\sqrt{10}\sqrt{72}$

13. $4\sqrt{5}\sqrt{10}$

14. $\sqrt{8}\sqrt{108}$

It should be noted that in order to for a radical to be an element of the set of real numbers, the value of the radicand must be either zero or positive, because squaring a negative term always results in a positive product. When a negative radicand exists, the term falls within the realm of the imaginary number system. Such discussion is beyond the scope of this text.

Multiplying Algebraic Fractions

Algebraic fractions are fractions that have a variable in the numerator, the denominator or both. Examples of algebraic fractions follow:

$$\frac{-2}{x} \qquad \frac{x}{5} \qquad \frac{6}{x+3} \qquad \frac{x+5}{x-1} \qquad \frac{2x+4}{3x-7} \qquad \frac{x^2+5x+6}{x^2+7x+10}$$

Multiplying simple fractions containing only monomials in numerator and denominator is as easy as multiplying numerical fractions: **multiply** numerators using Multiplication Rule # 1 or Multiplication Rule # 2 to obtain a product numerator, and **multiply** denominators to obtain a new product denominator. So,

$$\frac{-2}{x^2} \cdot \frac{x}{5} = \frac{-2x}{5x^2}$$

Recall, however, that as with multiplying numerical fractions, common factors in numerator and denominator cancel by the third usage of the multiplicative identity property. The cancellation is always one-to-one. Therefore it is better to reduce one-to-one first and then to multiply:

$$\frac{-2}{x^2} \cdot \frac{x}{5} = \frac{-2}{5x}$$

Try the following problems.

1. $\dfrac{2x}{5} \cdot \dfrac{-3}{4}$

2. $\dfrac{-5x}{6x} \cdot \dfrac{12x}{5x^2}$

3. $\dfrac{8x^2}{5x} \cdot \dfrac{30x^2}{4x^4}$

4. $\dfrac{3x}{7x} \cdot \dfrac{21}{6x}$

5. $\dfrac{-6x}{5x} \cdot \dfrac{15x}{48}$

6. $\dfrac{-42x}{36x} \cdot \dfrac{-6x}{14}$

7. $\dfrac{21}{x^3} \cdot \dfrac{-3x}{441}$

8. $\dfrac{169x}{64x} \cdot \dfrac{16x}{13}$

9. $\dfrac{8x}{12x} \cdot \dfrac{-3x}{32}$

10. $\dfrac{-24}{15x} \cdot \dfrac{75x}{576}$

Consider

$$\dfrac{-3}{-2x+3} \cdot \dfrac{-2x+3}{-6}$$

As with all operations involving algebraic fractions, the entire numerator and the entire denominator must ALWAYS be treated as binomials or polynomials surrounded by parentheses; it helps to put parentheses around

89

the expression. As with numerical fractions and the algebraic fractions in the previous subsection, sometimes a **common binomial factor** can be reduced from numerator and denominator before performing multiplication.

$$\frac{\overset{1}{\cancel{-3}}}{\underset{1}{(\cancel{-2x+3})}} \cdot \frac{\overset{1}{(\cancel{-2x+3})}}{\underset{2}{\cancel{-6}}} = \frac{1}{2}$$

Finding a common binomial factor in numerator and denominator may require factoring by using the distributive property in reverse. Thus,

$$\frac{-2}{x} \cdot \frac{(-2x+3)}{4x-6} = \frac{\overset{1}{\cancel{-2}}}{x} \cdot \frac{\overset{1}{(\cancel{-2x+3})}}{\underset{1}{-\underset{1}{\cancel{2}}(\cancel{-2x+3})}} = \frac{1}{x}$$

An all-too-common **error** in reducing fractions is

$$\frac{\cancel{-2}x+3}{\cancel{-2}} = x+3$$

This is the equivalent of incomplete division of, for example, 52 / 4.

$$\frac{\cancel{50}+2}{\cancel{4}} = 12.5 + 2$$

Instead, fractions can be reduced before multiplication if a common factor can be removed from all monomials in a numerator and denominator. The process is equivalent to using the distributive property in reverse.

$$\frac{-6x+4}{5} \cdot \frac{9x}{2} = \frac{-2(3x-2)}{5} \cdot \frac{9x}{2} = \frac{-9x(3x-2)}{5}$$

On some rare occasions, there is no common factor between either of the numerators and either of the denominators. Simply multiply.

11. $\dfrac{-5}{x} \cdot \dfrac{-2}{4x-16}$

12. $\dfrac{-7}{x} \cdot \dfrac{13x-26}{2x-4}$

90

13. $\dfrac{-24}{x} \cdot \dfrac{x+3}{2x+6}$

14. $\dfrac{-5x}{x} \cdot \dfrac{-2}{5x-20}$

15. $\dfrac{8}{x} \cdot \dfrac{-2}{8x+24}$

16. $\dfrac{-2}{x} \cdot \dfrac{(-2x+3)}{4x-6}$

17. $\dfrac{-5}{x} \cdot \dfrac{(-8x^2+4x)}{4x-2}$

18. $\dfrac{(x+3)(x-3)}{16} \cdot \dfrac{-32x}{2x-6}$

19. $\dfrac{8x^2(x+6)}{(x+3)(x+6)} \cdot \dfrac{x+3}{(x+7)(x+5)}$

20. $\dfrac{2x-6}{4x+8} \cdot \dfrac{6x-12}{5x-15}$

Again we see the necessity of using other operations in order to effect one of them; notice that multiplication of any term other than a monomial to another monomial necessitates the use of the distributive property of multiplication over addition.

DIVISION

Division is the inverse operation of multiplication. Reread the previous sentence: *division is the inverse operation of multiplication.* We can take any term, multiply it by any quantity, divide the result by that same quantity, and obtain the original term.

$$3(4) = 12 \qquad 12 \div 4 = 3$$

The Vocabulary of Division

$$\frac{\text{numerator}}{\text{denominator}} = \text{quotient} \qquad \text{divisor} \,|\, \overline{\text{dividend}}^{\text{ quotient}}$$

$$\text{dividend/divisor} = \text{quotient} \qquad \text{dividend} \div \text{divisor} = \text{quotient}$$

dividend - the term being divided. When division is expressed as a fraction, the dividend is called the **numerator**.
divisor - the number of equal pieces into which the dividend or numerator is being cut. When division is expressed as a fraction, the divisor is called the **denominator**.

rational number or term - a number or term expressed as a ratio, or comparison, of a/b where $b \neq 0$ (where b is not equal to zero)

Notice that the symbol, $\div$, is simply a fraction bar with dots above and below, whereby the dot above represents the numerator or dividend and the dot below represents the denominator or divisor. Students who are habituated to using this symbol should rehabituate themselves to write division as a fraction.

$$45 \div 3 = \frac{45}{3} = 15$$

Often, the slash, /, is used as a fraction bar, with parentheses around numerator and denominator, because it is much faster to type problems in this fashion.

In the same way that subtraction is addition of the subtrahend's additive inverse, division is multiplication by the divisor/denominator's multiplicative inverse. This idea is worth repeating: *division is multiplication by the divisor/denominator's multiplicative inverse.*

If multiplication is repeated addition, is division repeated subtraction?

$$\frac{30}{6} = 5, \quad 30 - 5 - 5 - 5 - 5 - 5 - 5 = 0$$

Given that five appears six times in a subtraction chain and the subtraction results in zero, it appears to be the case. Division, however, is more than that because, as with multiplication, *division usually changes the nature of the terms being divided.* Distance divided by time is a rate which produces velocity (speed), a completely different kind of term from the original two.

Factoring will also be treated in this chapter because it, like division, is the other inverse operation of multiplication and requires the same process as division. The difference between division and factoring is that in division, the dividend and divisor vanish to leave a quotient, whereas in factoring, the divisor remains as a factor, and the resulting quotient is the other factor. This idea can be expressed as follows:

$$\text{dividend} = (\text{divisor})(\text{quotient})$$
$$\text{product} = (\text{factor})(\text{factor})$$

The exception to the idea that factoring requires thinking in terms of division is when factoring second degree trinomials; in that instance, the use of addition and multiplication in the context of thinking backwards is required.

It is critical to understand that division by zero results in an undefined expression in mathematics. It simply makes no practical sense to divide anything by zero. Division by one, by the second usage of the multiplicative identity property, results in the same term; to divide by zero, then, results in no numerical answer. Analytically, division by zero is interpreted to mean an infinitely large number of undefined value. This will be seen when exploring the slope of a line in the chapter on applications. Therefore, it is crucial to insure that no denominator or divisor ever has a value of zero. Restrictions are placed on permissible variable values when division of algebraic expressions is involved and is explored in higher level algebra courses.

The rules for dividing all things mathematical are identical to the rules for multiplying all things mathematical.

Division Rule # 1

(1)   When dividend and divisor, or numerator and denominator, have the **SAME signs**, their quotient is **POSITIVE**.

Division Rule # 2

(2)   When dividend and divisor, or numerator and denominator, have **DIFFERENT signs**, their quotient is **NEGATIVE**.

Division Rules # 1 and # 2 Explained

The same table used to illustrate the multiplication rules works as well for the division rules. Each row of the division table corresponds to each row of the multiplication table. The division table was produced by reading each row of the multiplication table from right to left to perform its inverse.

Multiplication	Division
+ x + = +	+ ÷ + = +
+ x − = −	− ÷ − = +
− x + = −	− ÷ + = −
− x − = +	+ ÷ − = −

Dividing Integers

Division of integers follows the same two rules as multiplication of integers, that is:
(1)   When dividend and divisor, or numerator and denominator, have the **SAME signs**, their quotient is **POSITIVE**.
(2)   When dividend and divisor, or numerator and denominator, have **DIFFERENT signs**, their quotient is **NEGATIVE**.

1. $\dfrac{64}{-4}$ 3. $\dfrac{-52}{13}$

2. $\dfrac{-36}{-3}$ 4. $\dfrac{-51}{-3}$

94

5. $\dfrac{-42}{3}$ 11. $\dfrac{-112}{8}$

6. $\dfrac{60}{-4}$ 12. $\dfrac{117}{-9}$

7. $\dfrac{-105}{21}$ 13. $\dfrac{-256}{-16}$

8. $\dfrac{-92}{23}$ 14. $\dfrac{-324}{18}$

9. $\dfrac{-54}{3}$ 15. $\dfrac{289}{-17}$

10. $\dfrac{-96}{-16}$ 16. $\dfrac{-529}{-23}$

Dividing Numerical Fractions

The Multiplicative Inverse Property

$a \cdot \dfrac{1}{a} = 1$ any number or term times its multiplicative inverse equals one.

From this it follows that

$\dfrac{a}{b} \cdot \dfrac{b}{a} = 1$ any rational number or term times its multiplicative inverse equals one.

and

$\dfrac{a}{b} \div \dfrac{c}{d} = \dfrac{a}{b} \cdot \dfrac{d}{c}$ any fraction divided by another equals the by the reciprocal of the second fraction.

In the same way that to subtract means to add the subtrahend's additive inverse, to divide means to multiply the dividend by the divisor's multiplicative inverse. This is the principle behind, the essence of, division of fractions: performing the inverse of the inverse preserves the operation.

It is of highest importance that the student now distinguish between the additive inverse and the multiplicative inverse. Taking the additive inverse of a term involves moving horizontally toward zero and then beyond zero the same distance on the opposite side of the number line. Thus, a negative term becomes positive and a positive term becomes negative, but maintains the same absolute value, or distance from zero. Contrast taking the multiplicative inverse: the term remains on the same side of zero on the number line, whether negative or positive, but now executes a vertical flip, a head stand, whereby the numerator becomes the denominator and the denominator becomes the numerator. Its absolute value, or distance from zero, changes even though it remains on the same side of zero on the number line. The absorption of this concept is critical not just for this chapter but when isolating x in the chapters on applications.

$$\frac{4}{7} \div \frac{6}{14} = \frac{\overset{2}{\cancel{4}}}{\cancel{7}_{1}} \cdot \frac{\overset{2}{\cancel{14}}}{\cancel{6}_{3}} = \frac{4}{3}$$

Notice that the reciprocal of the second fraction was taken when the problem was converted to multiplication. Factors in numerator and denominator were found along the diagonals such that both parts of both fractions were reduced. Reduced numerators were multiplied to render the product numerator, and reduced denominators were multiplied to render the product denominator.

In some of the following problems, a perfect square and its root will cancel from numerator and denominator along a diagonal path once the problem has been converted into multiplication of the second fraction. This has been done to reiterate the notion that finding the root of a perfect square is division by its repeating factor in order to produce itself, as well as to reinforce the memorization of perfect squares.

1. $\dfrac{18}{4} \div \dfrac{-54}{10}$

2. $\dfrac{-32}{18} \div \dfrac{8}{3}$

3. $\dfrac{-5}{6} \div \dfrac{-15}{12}$

4. $\dfrac{-48}{64} \div \dfrac{-12}{16}$

5. $\dfrac{13}{23} \div \dfrac{65}{299}$

6. $\dfrac{14}{5} \div \dfrac{-210}{25}$

7. $\dfrac{-11}{24} \div \dfrac{-297}{6}$

8. $\dfrac{4}{132} \div \dfrac{6}{11}$

9. $\dfrac{-4}{182} \div \dfrac{10}{13}$

10. $\dfrac{-3}{14} \div \dfrac{12}{196}$

11. $\dfrac{-8}{361} \div \dfrac{2}{19}$

12. $\dfrac{-9}{21} \div \dfrac{-3}{441}$

Dividing Decimals

To divide decimals, place the dividend under the division bar and the divisor to the left of it. Move the decimal of the divisor to the right as many places as necessary until it is at the end, and then move the decimal in the dividend the same number of places. Place the decimal directly above the new position of the dividend's decimal and then divide using the following procedure.

Division of multi-digit numbers beyond those memorized in multiplication tables is known as long division. This requires the use of three skills: estimation, multiplication and subtraction. Consider

$$\overline{.234 \mid 30.10176}$$

The divisor, in this case, .234, first must be converted to a whole number. This is accomplished by moving the decimal as many places as necessary until the end, in this case three; in turn, the decimal in 30.10176, the dividend, must be moved the same number of places to the right.

```
          128.64
234. | 30101.76
     - 234
       670
     - 468
       2021
     - 1872
       1497
     -  1404
         936
     -   936
           0
```

Estimation is the first step; the student must decide how many times 234 goes into 301 without going over 301. Some trial and error by means of multiplication is often necessary. When 1 is established as the number, 1 is *multiplied* to 234 to produce 234; this product is *subtracted* from 301, leaving a remainder of 67. The next digit in the dividend, in this case 0, is brought down and the estimation begins again. The process repeats until (a) a terminating or repeating decimal is obtained, (b) the problem indicates how many decimal places are required or (c) it becomes obvious that the quotient is an irrational number.

The process is decidedly tedious, frustrating even the strongest math students in elementary school. It is understandable that many students use calculators to complete worksheets on this topic. Yet, much is lost when students do not endure the tedium of manual computation. The first and best exposure to the skill of estimation is to be found in adequate practice in long division without a calculator; when a calculator is used, only the division button is pressed, and students miss the understanding that division is dependent upon estimation, multiplication and subtraction.

Try the following problems without a calculator.

1.
$$.61 \mid \overline{1.50792}$$

2.
$$.56 \mid \overline{19.7344}$$

3. $.27 \; | \; \overline{116.64}$ 5. $.483 \; | \; \overline{21.4935}$

4. $7.92 \; | \; \overline{425.304}$ 6. $1.86 \; | \; \overline{1185.3408}$

Dividing Numbers Expressed in Scientific Notation

In the same way that multiplication is repeated addition, division is repeated subtraction. Hence, it should come as no surprise that, whereas multiplying numbers expressed in scientific notation follows the procedure multiply–keep – add, dividing numbers expressed in scientific notation follows the procedure **divide** coefficients using Division Rule # 1 or # 2– **keep** the base ten – **subtract** the exponents by adding the opposite and using Addition Rule #1 or #2.

Ex. $(4.5 \times 10^5) \div (- 1.5 \times 10^0) = (4.5/- 1.5) \times 10^{5-0} = \; \textbf{- 3.0 x 10}^{\textbf{5}}$

1. $(- 1.36 \times 10^6) \div (- 1.7 \times 10^4)$

2. $- 1.36 \times 10^6 \div (- 1.7 \times 10^{-8})$

3. $- 6.867 \times 10^{-5} \div - 1.09 \times 10^{-3}$

4. $- 6.867 \times 10^2 \div - 1.09 \times 10^7$

5. $11.872 \times 10^5 \div - 4.48 \times 10^3$

6. $4.48 \times 10^5 \div - 8.96 \times 10^9$

7. $(- 3.2 \times 10^{-5}) \div (- 9.6 \times 10^3)$

8. $- 9.8 \times 10^{-5} \div - 2.2 \times 10^{12}$

9. $- 4.52466 \times 10^{-3} \div 9.31 \times 10^{-7}$

10. $- 8.48 \times 10^{-4} \div - 6.4 \times 10^{-6}$

Dividing Monomials

To divide algebraic monomials,

Divide coefficients – **Keep** the base – **Subtract** the exponents
using Division Rule   by adding the opposite and
1 or # 2   using Addition Rule # 1 or # 2

Consider

$$\frac{x^5}{x^3}$$

This is

$$\frac{x \bullet x \bullet x \bullet x \bullet x}{x \bullet x \bullet x}$$

By the third usage of the Multiplicative Identity Property, we can reduce factors in numerator one-to-one to obtain

$$\frac{x \bullet x \bullet x \bullet x \bullet x}{x \bullet x \bullet x} \ = \ x \bullet x = x^2$$

The rule which follows is to divide coefficients using Division Rule # 1 or # 2, to keep the base and to subtract the exponents by performing **keep-plus-opposite** and applying Addition Rule # 1 or # 2.

$$\frac{x^5}{x^3} = x^{5-3} \ = \ x^2$$

Now consider

$$\frac{x^5}{x^5}$$

This is

$$\frac{x \bullet x \bullet x \bullet x \bullet x}{x \bullet x \bullet x \bullet x \bullet x}$$

Reducing factors in numerator and denominator one-to-one, we obtain

$$\frac{x \bullet x \bullet x \bullet x \bullet x}{x \bullet x \bullet x \bullet x \bullet x} = 1$$

However, by applying the division rules for monomials,

$$\frac{x^5}{x^5} = x^{5-5} = x^0$$

When two things are equal to a third thing, the two things are equal to each other; this is known as the Transitive Property of Equality. Therefore,

$$x^0 = 1$$

Realize that we can do this with any base, numerical or variable.

$$\frac{2^4}{2^4} = \frac{2 \bullet 2 \bullet 2 \bullet 2}{2 \bullet 2 \bullet 2 \bullet 2} = 2^{4-4} = 2^0 = 1$$

When any power is divided by itself, the difference of the exponents will always equal zero and the cancellation of factors in numerator and denominator will always equal one. Therefore,

$$a^0 = 1, \, a \, \varepsilon \, \Re \text{ (any number or term to the zero power equals one, where}$$
$$a \text{ is an element of the real number system.)}$$

Now consider

$$\frac{x^3}{x^5}$$

This is

$$\frac{x \bullet x \bullet x}{x \bullet x \bullet x \bullet x \bullet x}$$

Reducing factors in numerator and denominator one-to-one, we obtain

$$\frac{x \bullet x \bullet x}{x \bullet x \bullet x \bullet x \bullet x} = \frac{1}{x^2}$$

If we use the division rules for monomials, we obtain

$$\frac{x^3}{x^5} = x^{3-5} = x^{-2}$$

Again, when two things are equal to a third thing, they are equal to each other. Therefore

$$\frac{1}{x^2} = x^{-2}$$

In words, a base raised to a negative exponent equals the reciprocal of the base (flipped upside down into the denominator with a numerator of one) with the exponent's sign changed to positive. Likewise, if a negative power appears in the denominator of a fraction, it can be moved into the numerator with the exponent changed to positive. Because positive exponents are preferred to negative exponents, it is preferable to convert all powers to have positive exponents, even if it means expressing powers as rational expressions. Consider the following problem where the division rules for monomials are used, and an alternative method of taking reciprocals of all powers with negative exponents as a first step, thereby changing those exponents to positive, is used:

Using Division Rules

$$\frac{x^{-5}y^2}{x^5y^{-3}} = x^{-5-5}y^{2--3} = x^{-10}y^5 = \frac{y^5}{x^{10}}$$

Making All Negative Exponents Positive
As a First Step

$$\frac{x^{-5}y^2}{x^5y^{-3}} = \frac{y^2y^3}{x^5x^5} = \frac{y^{2+3}}{x^{5+5}} = \frac{y^5}{x^{10}}$$

It is important to remember that when using the division rules, a power with a negative exponent remains in the numerator until its reciprocal is taken and its exponent is made positive. It is equally important to remember that converting all powers to have positive exponents as a first step is not always convenient, as the student may still have to subtract exponents.

Using Division Rules

$$\frac{x^2 x^{-5} y^2}{x^5 y^{-3}} = x^{2+(-5)-5} y^{2--3} = x^{-8} y^5 = \frac{y^5}{x^8}$$

Making All Negative Exponents Positive
As A First Step

$$\frac{x^2 x^{-5} y^2}{x^5 y^{-3}} = \frac{x^2 y^2 y^3}{x^5 x^5} = \frac{x^2 y^{2+3}}{x^{5+5}} = x^{-8} y^5 = \frac{y^5}{x^8}$$

Recall that a monomial is a multiplication chain and that the multiplication chain may contain parenthetic binomials or other polynomials inside parentheses. The rules for division apply to the powers, visible or invisible, attached to their parentheses.

$$\frac{x^5 (x+y)^3}{x^9 (x+y)^4} = x^{5-9}(x+y)^{3-4} = x^{-4}(x+y)^{-1} = \frac{1}{x^4(x+y)}$$

Try the following problems using the two rules of division.

1. $\dfrac{-12x^6}{3x^2}$

2. $\dfrac{-12x^3}{-3x^4}$

3. $\dfrac{-56x^8}{8x^7}$

4. $\dfrac{-x^{-2}}{-4x^2}$

5. $\dfrac{-56x^{-5}}{4x^{-2}}$

6. $\dfrac{-9.3x^{-8}y^4}{3x^9 y^5}$

7. $\dfrac{63x^{-4}y^{-7}}{-9x^{-2}y^3}$

8. $\dfrac{-x^5y^6(x+y)^5}{-3x^{-8}y^{-3}(x+y)^3}$

9. $\dfrac{4x^3y^7z^2(x+y)}{-3x^{-2}y^9z^5(x+y)}$

10. $\dfrac{-11x^{-11}y^8z(x+y)^{-1}}{44x^{-3}y^9z^3(x+y)^{-2}}$

Factoring Binomials

As stated in the chapter on addition, once an addition chain polynomial is formed, it becomes a unity which can only be dissolved by means of factoring or division. In this and the following subchapters, we will explore the methods by which addition chain polynomials can be undone. Factoring, the other inverse operation of multiplication, is a process of using division to think backwards while applying the distributive property in reverse. To factor $3x^2 + 12x$, one must identify the greatest common factor in both monomials of the binomial. In this case, the greatest common numerical factor is 3, and the greatest common algebraic factor is x; together, the greatest common factor is 3x. Pull a factor of three-x out of each term in front of a set of parentheses, and write the remaining factors inside of the parentheses:

$3x^2 + 12x$ factored is $3x(x + 4)$
$5x + 25$ factored is $5(x + 5)$

To check if you have factored correctly, mentally expand forward using the distributive property to insure that you obtain the original binomial.

It is important to realize that factoring a negative from both terms of the binomials will result in a plus sign inside of parentheses. So,

$$-3x^3 - 18x$$

becomes

$$-3x(x^2 + 6)$$

104

because expanding by use of the distributive property will reintroduce the minus sign inside of parentheses to produce the original binomial. All factoring should be checked, at least mentally, with a forward expansion using the distributive property to insure that the correct signs in front of each term are where they should be.

Sometimes it is useful to factor a negative from both terms even when the negative is visible in only the first term of the binomial. Thus,

$$-4x^2 + 14x$$

becomes

$$-2x(2x - 7)$$

1. $14x^2 - 42x$

2. $-25x + 45$

3. $-13x^2 + 52x$

4. $-18x^2 - 54x$

5. $-3x^2 - 45x$

6. $x^2 - x$

7. $17x^2 + 85x$

8. $6x^2 - 60x$

9. $-12x^2 + 132x$

10. $8x^2 - 64x$

A special form of polynomial is the difference of perfect squares binomial. This is not to be confused with the perfect square trinomial in the next subchapter. The difference of perfect squares is a second degree binomial comprised of an x-squared term, a minus sign and a numerical perfect square. In more complex forms, there may be two variables involved. Factoring these binomials is as simple as opening two sets of parentheses, inserting the square roots of the two terms, and placing a plus sign in between one and a minus sign in between the other. For example,

$$x^2 - 9 \text{ becomes } (x + 3)(x - 3)$$

Notice that when the binomials are expanded by FOIL, the outer and inner products add to zero, leaving a difference of squares. Geometrically, this can be represented as a rectangle with length and width represented by each linear binomial and area represented by the difference of squares. Also note that the resulting area is not itself a square because of the different values of each binomial.

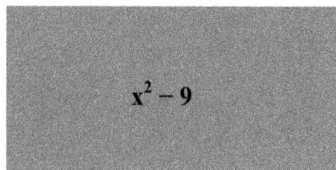

x– 3

$x^2 - 9$

x + 3

Factor these differences of squares.

11. $x^2 - 81$

12. $x^2 - 121$

13. $4x^2 - 64$

14. $9x^2 - 100$

15. $16x^2 - 144$

16. $25x^2 - 169$

17. $196 - 36x^2$

18. $225 - 49x^2$

19. $256x^2 - 400\, y^2$

20. $\dfrac{y^2}{289} - 484x^4$

Occasionally, factoring a binomial will result in a monomial factor multiplied to a remaining binomial which is a difference of perfect squares. Consider

$$16x^2 - 64$$

Factoring out the greatest common factor first, this becomes

$$= 16(x^2 - 4)$$

Factoring out the difference of squares second, the expression is factored completely.

$$= 16(x + 4)(x - 4)$$

21. $3x^2 - 27$

22. $4x^2 - 100$

23. $2x^2 - 98$

24. $5x^2 - 180$

25. $6x^2 - 600$

26. $4x^2 - 324$

27. $5x^2 - 605$

28. $3x^2 - 507$

29. $12x^2 - 1,728$

30. $16x^2 - 256y^2$

Dividing a Binomial by a Monomial

A binomial divided by a monomial should be thought of as a fraction which resulted from adding two like fractions with unlike numerators and being in need of further simplification. Thus

$$\frac{12x^7 - 6x^3}{3x^2}$$

was the result of adding some expression equivalent to

$$= \frac{12x^7}{3x^2} - \frac{6x^3}{3x^2}$$

It is clear that both fractions can be further simplified. Applying the **divide-keep-subtract** rules for division of monomials, we obtain

$$= 4x^5 - 2x$$

1. $\dfrac{15x^8 + 9x^4}{3x^3}$

2. $\dfrac{28x^{10} - 14x^3}{7x^2}$

3. $\dfrac{8x^4 - 16x^3}{8x^2}$

4. $\dfrac{-3.6x^4 - 2.4x^2}{1.2x^2}$

5. $\dfrac{56x^3 - 6x^2}{7x^2}$

6. $\dfrac{-5.4x^4y^3 - 6x^3y}{3x^2y}$

7. $\dfrac{x^{-3}y - 2x^2y}{2x^{-2}y}$

8. $\dfrac{1.6x^{-2}y^3 - 6x^{-3}y^4}{4x^2y^4}$

Factoring Trinomials

As stated in the chapter on addition, once an addition chain polynomial is formed, it becomes a unity which can only be dissolved by means of factoring or division. Consider the trinomial

$$x^2 + 5x + 6$$

We have seen in the chapter on multiplying binomials that this quadratic trinomial is the product of two linear binomial factors, $(x + 2)(x + 3)$.
Use of the distributive property twice, popularly known as FOIL, produces four products, two of which are like and combined into a single term.

	x	+ 2
x	Multiply - - Keep – Add x^2	Multiply - - Keep – Add $+ 2x$
+ 3	Multiply - - Keep – Add $+ 3x$	Multiply - - Keep – Add 6

5x

The three unlike monomials are then linked in an addition chain.

Therefore, simply unlinking the plus signs and separating the unlike monomials, an all-too-common error, is not the final step but only the first step toward reversion back to the original binomial terms.

	x	+ ?
x	Multiply - - Keep – Add x^2	Multiply - - Keep – Add $?x$
+ ?	Multiply - - Keep – Add $?x$	Multiply - - Keep – Add 6

5x

Whatever replaces the red and blue question marks must add up to 5 and multiply to 6. By guess-and-check, 2 and 3 can be found.

The factoring, or undoing, of a quadratic trinomial, like the undoing of exponentiation, is a process of thinking backwards. The common method is to open up two sets of parentheses and to insert a first degree x term in the first positions of each binomial. A simple process of guess-and-check ensues: what two numbers add (or subtract) to the coefficient of the x term, and what same two numbers multiply to the third constant term? These numbers *with their signs attached in front* are inserted into the parentheses. So, given $x^2 + 5x + 6$,

$$(x \qquad)(x \qquad)$$

$$\underline{\qquad} + \underline{\qquad} = 5$$

$$\underline{\qquad} \cdot \underline{\qquad} = 6$$

One and five add to six and multiply to five, the reverse of what we want, so these create the factors of a different trinomial. Two and three add to five and multiply to six; hence,

$$(x + 2)(x + 3)$$

is the factored form of $x^2 + 5x + 6$.

The process is only slightly trickier when different signed terms are involved. Consider

$$x^2 - x - 20$$

$$(x \qquad)(x \qquad)$$

Since every negative product results from one positive and one negative factor, one factor of negative twenty must be negative and the other must be positive. Therefore, when a negative and positive are combined, Addition Rule # 2 must be used to form the middle term, that is, they will *subtract* to the coefficient of x.

$$\underline{\quad-\quad} + \underline{\qquad} = -1$$

$$\underline{\quad-\quad} \cdot \underline{\qquad} = -20$$

110

Positive five and negative four subtract to positive one, so this is not the correct combination. However,

$$\underline{\quad -5 \quad} + \underline{\quad 4 \quad} = \; -1$$

$$\underline{\quad -5 \quad} \cdot \underline{\quad 4 \quad} = \; -20$$

Therefore,

$$(x-5)(x+4)$$

is the correct factored form of $x^2 - x - 20$.

Of course, if the student preferred using the matrix to multiply binomials, he may also prefer using it to factor trinomials back into linear binomials.

The format is different but the thinking is identical: to find two numbers which add to negative one and multiply to negative twenty. The numbers are still -5 and 4; the factored form is still expressed as $(x-5)(x+4)$.

When factoring a trinomial with a negative leading coefficient, that is, a negative coefficient attached to the highest power of x, which is the x-squared term here, one positive x and one negative x are placed in each set of parentheses or next to and above the upper left box of the matrix. Thus, to factor

$$-x^2 + 7x - 12$$

we must bear in mind that the number replacing the red question mark must be negative, because when it is multiplied to $-x$, the product becomes positive. Notice that two seemingly different pairs of binomials result:

$$-x \qquad + \quad ?$$

X	Multiply - - Keep − Add $-x^2$	Multiply - - Keep − Add $?x$
$+?$	Multiply - - Keep − Add $?x$	Multiply - - Keep − Add -12

$$7x$$

$$(-x+4)(x-3) \ \text{ or } \ (-x+3)(x-4)$$

For students who know that the next and last step is to set each binomial equal to zero in order to find the solutions, reserved in this text for the last chapter, we realize that no matter which correct pair of binomials result from your guess-and-check, the correct solutions are $x = 3$ and $x = 4$.

Correct Pair		Correct Pair	
$-x+4=0$ $x=4$	$x-3=0$ $x=3$	$-x+3=0$ $x=3$	$x-4=0$ $x=4$
Incorrect Pair		Incorrect Pair	
$-x-4=0$ $x=-4$	$x+3=0$ $x=-3$	$-x-3=0$ $x=-3$	$x+4=0$ $x=-4$

Needless to say, this work performed without a proper check is tedious and useless; the only way to know whether you have found a correct pair of binomials is to multiply them to obtain the original trinomial.

$(-x + 4)(x - 3)$
$= -x^2 + 4x - 3(-x) - 12$
$= -x^2 + 4x + 3x - 12$
$= -x^2 + 7x - 12$ ✓

$(-x + 3)(x - 4)$
$= -x^2 + 3x - 4(-x) - 12$
$= -x^2 + 3x + 4x - 12$
$= -x^2 + 7x - 12$ ✓

When the lead coefficient, that is, the coefficient of the x-squared term, is two or greater, the "eyeglass" method is used.

$3x^2 + 10x + 8$ — Create a new temporary quadratic: multiply the lead coefficient by the constant term for a new constant term; make the lead coefficient one.

$x^2 + 10x + 24$
$(x + 6)(x + 4)$
$(3x + 6)(3x + 4)$
$(x + 2)(3x + 4)$

Factor using the ordinary guess-and-check procedure. Reintroduce the original lead coefficient to both x terms. Reduce the binomial in which both monomials are divisible by the lead coefficient; keep the other binomial intact. If both binomials are divisible by *a factor of* the lead coefficient, reduce both binomials.

$x = 2 \,|\, x = -\frac{4}{3}$ — The check is left to the student.

1. $x^2 + 6x + 5$

2. $x^2 + 9x + 8$

3. $x^2 + 7x + 12$

4. $x^2 - 7x + 12$

5. $x^2 - x - 12$

6. $x^2 - 5x + 6$

7. $x^2 - 5x - 36$

8. $x^2 - 16x + 48$

9. $x^2 - 32x - 144$

10. $x^2 - 12x - 64$

11. $-x^2 - 7x - 12$

12. $-x^2 - 10x - 21$

113

13. $-x^2 + 16x - 55$

14. $-x^2 + 15x - 36$

15. $2x^2 + 11x - 21$

16. $2x^2 - 8x - 10$

17. $3x^2 - 11x - 10$

18. $3x^2 + 21x + 18$

There is a special form of trinomial, the perfect square trinomial, not to be confused with the difference of perfect squares *binomial* discussed in the previous section and so called because of its corresponding geometric representation.

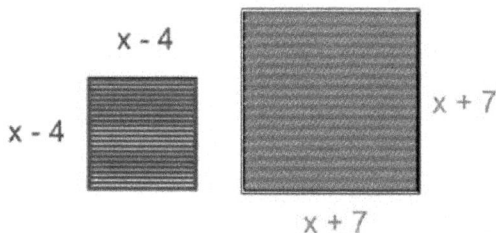

Notice that the multiplication of a linear term to itself, in this case a binomial, has the geometric effect of creating a square whose area is the product of the side repeated. Expanding $(x - 4)^2$ and $(x + 7)^2$ using FOIL produces $x^2 - 8x + 16$ and $x^2 + 14x + 49$, respectively.

Hence, to factor a perfect square trinomial, the process is this simple: take one-half of the coefficient of the x term together with its sign and place it into parentheses with a first degree x term in front. Factoring the previous two perfect square trinomials, $x^2 - 8x + 16 = (x - 4)^2$ and $x^2 + 14x + 49 = (x +7)^2$. Naturally, one will recognize a perfect square trinomial when taking *one-half of the coefficient of x and squaring it* produces the constant term in the trinomial, in the cases above, 16 and 49, respectively. This completes the process for factoring perfect square trinomials.

19. $x^2 + 16x + 64$

114

20. $x^2 - 20x + 100$

21. $x^2 - 42x + 441$

22. $x^2 - 46x + 529$

23. $x^2 + 48x + 576$

24. $x^2 - 28x + 196$

25. $x^2 + 38x + 361$

26. $x^2 + 36x + 324$

27. $x^2 + 44x + 484$

28. $x^2 - 50x + 625$

Factoring and Dividing Radicals

You already have had much practice in factoring numerical and algebraic radicals. To review, converting radicals into simplest form involves pulling out a factor of the largest possible perfect square by taking its square root, leaving the smallest radicand under the radical. The key to simplifying radicals is, by trial and error, to divide the radicand by 2, 3, 5, 6, or 7. One of these numbers will be the remaining radicand; the other number will be the largest possible perfect square whose square root can then be taken and converted into an integral factor, placed in front of the radical and creating a multiplication chain. Try dividing by two first; if the other number is not a perfect square, try to divide the radicand by three, and so on, until a perfect square is obtained.

In the same way that a radical of factors may become a factor of radicals and vice versa, so too a radical of a quotient may become a quotient of radicals; if the radicals are perfect squares, they reduce to a simple fraction.

$$\sqrt{\frac{1}{4}} = \frac{\sqrt{1}}{\sqrt{4}} = \frac{1}{2}$$

Otherwise, numerator, denominator or both may remain as radicals.

$$\sqrt{\frac{1}{2}} = \frac{\sqrt{1}}{\sqrt{2}} = \frac{1}{\sqrt{2}}$$

Ex. 1.   Simplify $\dfrac{\sqrt{4x^5}}{\sqrt{288x^3}}$

Solution: We make the quotient of radicals the radical of a quotient.

$$= \sqrt{\frac{4x^5}{288x^3}}$$

Reduce/divide the coefficients, **keep** the variable base and **subtract** the exponents.

$$= \sqrt{\frac{x^2}{72}}$$

Simplify the radical.

$$= \frac{x}{6\sqrt{2}}$$

Ex. 2.

$$\frac{\sqrt{24} + 2\sqrt{72}}{2}$$

Simplify radicals in the binomial numerator.

$$= \frac{2\sqrt{6} + 12\sqrt{2}}{2}$$

Reduce the coefficients; this can be done by separating the binomial into two fractions with the same denominator.

$$= \frac{2\sqrt{6}}{2} + \frac{12\sqrt{2}}{2}$$

In time, you will be able to skip this step and go to the last.

$$= \sqrt{6} + 6\sqrt{2}$$

116

1. $\dfrac{\sqrt{3}}{\sqrt{48}}$

3. $\dfrac{\sqrt{6}}{\sqrt{54}}$

2. $\dfrac{\sqrt{2}}{\sqrt{128}}$

4. $\dfrac{\sqrt{98}}{\sqrt{2}}$

5. $\dfrac{\sqrt{192}}{\sqrt{3}}$

10. $\dfrac{\sqrt{363x^4}}{\sqrt{3x^5}}$

6. $\dfrac{\sqrt{4}}{\sqrt{144}}$

11. $\dfrac{\sqrt{507x^7}}{\sqrt{3x^6}}$

7. $\dfrac{\sqrt{125}}{\sqrt{5}}$

12. $\dfrac{12+\sqrt{52}}{2}$

8. $\dfrac{\sqrt{72}}{\sqrt{2}}$

13. $\dfrac{25-\sqrt{75}}{15}$

9. $\dfrac{\sqrt{2x^6}}{\sqrt{512x^4}}$

14. $\dfrac{-24-2\sqrt{72}}{3}$

Factoring and Dividing Algebraic Fractions

In this subsection we see that division of algebraic fractions is dependent upon factoring; factoring, in turn, is dependent upon addition, subtraction and multiplication. It is in these problems where the all-too-common error of unhinging the addition or subtraction links is attempted and necessarily fails. To divide algebraic fractions, multiply the first fraction by the reciprocal of the second fraction.

Ex. 1. $\dfrac{x^2+10x+24}{x^2-9x+18} \div \dfrac{x^2-x-20}{x^2+2x-15}$

117

Take the reciprocal of the second fraction; replace ÷ with ·.

$$= \frac{x^2+10x+24}{x^2-9x+18} \cdot \frac{x^2+2x-15}{x^2-x-20}$$

Factor both numerators and both denominators; eventually, you will be able to do the first two steps in one step.

$$= \frac{(x+6)(x+4)}{(x-3)(x-6)} \cdot \frac{(x+5)(x-3)}{(x-5)(x+4)}$$

Reduce binomial factors from numerators and denominators; examining a diagonal path between fractions may be necessary.

$$= \frac{(x+6)(x+5)}{(x-6)(x-5)}$$

Leave the simplified fraction in factored form.

1. $\dfrac{x^2-7x+12}{x^2+2x-15} \div \dfrac{x^2-6x+8}{x^2+7x+10}$

2. $\dfrac{x^2-x-42}{x^2-x-6} \div \dfrac{x^2-16x+63}{x^2-12x+27}$

3. $\dfrac{x^2+3x-28}{x^2-5x+4} \div \dfrac{x^2+9x+14}{x^2-1}$

4. $\dfrac{x^2+10x-11}{x^2+5x-6} \div \dfrac{x^2+8x-33}{x^2+2x-24}$

5. $\dfrac{x^2-17x+66}{x^2-4x-5} \div \dfrac{x^2-5x-66}{x^2+x-30}$

6. $\dfrac{x^2+11x-26}{x^2-10x+16} \div \dfrac{x^2+12x-13}{x^2-9x+8}$

7. $\dfrac{2x^2-x-1}{x^2+3x+4} \div \dfrac{2x^2-9x-5}{x^2-16}$

8. $\dfrac{2x^2+5x+3}{x^2+11x-12} \div \dfrac{2x^2-x-6}{x^2+10x-24}$

9. $\dfrac{5x^2-x-4}{x^2+x-2} \div \dfrac{5x^2+19x+12}{x^2-x-6}$

10. $\dfrac{3x+9}{x^2+12x+20} \div \dfrac{9x+27}{x^2+8x+12}$

ORDER OF OPERATIONS

In this chapter we investigate the order in which numerical and algebraic expressions are evaluated based upon the operations involved in those expressions.

Order of Operations

The operations in any numerical or algebraic expression or equation must be evaluated in the following order:

1. Expressions involving operations inside **parentheses** are protected.
    In a fraction, the **fraction bar** acts as protecting the entire numerator and the entire denominator with their respective sets of parentheses.
    Absolute value bars likewise act as parentheses protecting the entire expression therein. Expressions in all such operators are evaluated first.
2. **Exponents** and **radicals** are evaluated from left to right.
3. **Multiplication** and **division** are evaluated from left to right.
4. **Addition** and **subtraction** are evaluated from left to right.

Evaluating Numerical Expressions

Ex. 1.
$$2(3) + \frac{8}{4} - 5(4)$$

The operations involved are addition, subtraction, multiplication and division; there are no exponents or parentheses to consider. Following the order of operations, multiplication and division are performed first from left to right.

$$= 6 + 2 - 20$$

Next, addition and subtraction are performed from left to right.

$$= 8 - 20 = -12$$

119

First the multiplication/division chains are evaluated; then, the resulting addition/subtraction chain is evaluated from left to right.

Ex. 2. $\qquad 2(3)^2 + \dfrac{8}{4} - 5(4)$

An exponent indicating repeated multiplication is present in this expression; the power is evaluated first.

$$= 2(9) + \dfrac{8}{4} - 5(4)$$

It is immaterial that the exponent appears after ordinary multiplication in this three-link multiplication chain; *in any multiplication chain*, **repeated** *multiplication is always performed* **before ordinary** *multiplication* even though in most cases it appears after ordinary multiplication in the chain. Evaluating multiplication and division from left to right, followed by addition and subtraction from left to right, we obtain

$$= 18 + 2 - 20 \quad = \quad 20 - 20 \quad = \quad 0$$

Ex. 3. $\qquad 2^3 3^2 + \dfrac{8}{4} - 5(4)$

The first multiplication chain, longer than before, contains two repeated multiplication links; those are performed first. Two-cubed is eight and three-squared is nine:

$$= 8(9) + \dfrac{8}{4} - 5(4)$$

$$= 72 + 2 - 20 = 54$$

Nor does it matter where the multiplication/repeated multiplication chain occurs in the addition chain; repeated multiplication, within or apart from a multiplication chain, is always performed first, as the next example illustrates.

Ex. 4. $\quad \dfrac{15}{5} + 2^3 3^2 - 5(4)$ $\qquad$ Evaluate powers.

$$= 3 + (8)(9) - 20 \qquad \text{Evaluate the multiplication chain.}$$
$$= 3 + 72 - 20 \qquad \text{Evaluate the addition/subtraction chain.}$$
$$= 55$$

Note that the division and multiplication operations were evaluated simultaneously with the powers. This simply consolidated two steps into a single step. As a technical matter, it is acceptable to evaluate the separate multiplication/division chains simultaneously, provided that the outcome does not contradict the correct order of operations.

Because the inverse of exponentiation is finding the corresponding root, radicals share with exponentiation the same level of priority in the order of operations.

Ex. 5. $5\sqrt{49} - 3^3 5^2$ Evaluate powers and radicals.
$$= 5(7) - (27)(25) \qquad \text{Evaluate the multiplication chains.}$$
$$= 35 - 675 \qquad \text{Evaluate the subtraction chain.}$$
$$= -640$$

When addition, subtraction or some combination of operations must be given the highest priority, they are elevated to highest priority by placing parentheses around the terms involved.

Ex. 6. $(2^3 - 3^2)^4 + (3 - 4 + 2)$ Evaluate powers inside parentheses.
$$= (8 - 9)^4 + (3 - 4 + 2) \qquad \text{Evaluate subtraction inside parentheses.}$$
$$= (-1)^4 + (1) \qquad \text{Evaluate power.}$$
$$= 1 + (1) \qquad \text{Evaluate addition chain.}$$
$$= 2$$

In the case of a negative base raised to an even power, the student must distinguish between

$$(-2)^4 \qquad \text{and} \quad -2^4$$

In the first case, the exponent is attached to the parentheses and is thus obligated to power up the entire expression inside parentheses.

$$(-2)^4 = (-2)(-2)(-2)(-2) = 16$$

121

In the second case, the exponent is attached only to the number two. The negative symbol is interpreted to be a factor of negative invisible one. Since multiplication is performed after repeated multiplication, the result is

$$-2^4 = -1 \cdot 2^4 = -1 \cdot 16 = -16$$

In the case of a negative base raised to an odd power, the outcome will be the same whether the base with the negative sign attached is inside parentheses or not, because a negative base raised to an odd power is always negative.

$$(-2)^3 = (-2)(-2)(-2) = -8$$
$$-2^3 = -1 \cdot 2^3 = -1 \cdot 8 = -8$$

Note the application of this distinction in the following problem.

Ex. 7.　$4\sqrt{169} - (-2^4 - 5)$　　Evaluate power inside parentheses.
　　　$= 4\sqrt{169} - (-16 - 5)$　　Evaluate subtraction inside parentheses;
　　　　　　　　　　　　　　　　　the root may simultaneously be evaluated.
　　　$= 4(13) - (-21)$　　　　　Evaluate multiplication chain.
　　　$= 52 + 21$　　　　　　　　Evaluate addition chain.
　　　$= 73$

The order of operations applies to combinations of operations inside parentheses or brackets. When there are two or more sets of parentheses within parentheses, the expressions within the innermost set of parentheses is evaluated first.

Ex. 8.　$(4 - 3^2 - (9 - 13)^3 + 15/5) + 24$　Evaluate innermost parentheses.
　　　$= (4 - 3^2 - (-4)^3 + 15/5) + 24$　Evaluate powers inside parentheses.
　　　$= (4 - 9 - (-64) + 15/5) + 24$　Evaluate division in parentheses.
　　　$= (4 - 9 - (-64) + 3) + 24$　Evaluate chain in parentheses.
　　　$= (-5 + 64 + 3) + 24$　　Evaluate addition.
　　　$= 62 + 24$
　　　$= 86$

1.　　$3 + 4^2$
2.　　$8 - 4(6 + 7)^2$
3.　　$9 - 5(8 - 14 + 3)^3$
4.　　$3^4 - 5^2$
5.　　$2^4(3 - 10)^2$

6.　　$\dfrac{3^2 + 5^2}{2} - \dfrac{6^2 - 2^3}{4}$

122

7. $5\sqrt{121} - 3^4(-2^3(2-7) + 16/4)$
8. $6\sqrt{100} - 3^4[(-2)^3(2-7)]$
9. $-3[16/4 - 2(3^2 - 2^5(4 - 2(3)) - 24)(-2)]$

Evaluating Algebraic Expressions

Consider

$$\frac{x}{y} + x^3y^2 - xy$$

The order of operations is followed when numerical substitutions are made into algebraic expressions. If $x = 2$ and $y = 4$, then substitution yields

$$\frac{2}{3} + 2^3\,4^2 - 2(4) = \frac{2}{3} + 8(16) - 8 = 120\,\tfrac{2}{3}$$

When a variable is the exponent of a numerical (constant) base, we have what is known as an *exponential function*. The expression 2^x is evaluated with the same level of priority as other powers.

When $x = -3$, then $\quad 2^{-3} = \dfrac{1}{2\cdot 2\cdot 2} \quad = \frac{1}{8}$

when $x = -2$, then $\quad 2^{-2} = \dfrac{1}{2\cdot 2} \quad = \frac{1}{4}$

when $x = -1$, then $\quad 2^{-1} = \dfrac{1}{2}$

when $x = 0$, then $2^0 = 1$
when $x = 1$, then $2^1 = 2$
when $x = 2$, then $2^2 = 4$
when $x = 3$, then $2^3 = 8$

1. $A_{trapezoid} = \dfrac{(b_1 + b_2)\,h}{2}$, $b_1 = 16$, $b_2 = 10$, $h = 4$

2. $\dfrac{x^2 - x - 2}{x - 1}$, $\quad x = -2$

123

3. $\dfrac{x^2 - x - 2}{x - 1}$, $x = 1$

4. $(x + 3)(y + 2)$, $x = 2, y = 6$

5. $\dfrac{4}{(x - 1)^2}$, $x = -3$

6. $2x^3 - 4x^2 - 15x + 5$, $x = -4$

7. $x^4 - 8x^3y + 24x^2y^2 - 32xy^3 + 16y^4$, $x = 2, y = -3$

8. $2^x \sqrt{x^2}(3^2 - 43y^3 - 4\sqrt{25} + 120/5)(3 - 1 - 7)$, $x = 9, y = -1$

APPLICATIONS IN ONE VARIABLE:
X AS AN UNKNOWN

Algebra is the mathematics of solving for unknown values represented by variables. In the first five chapters of this text, we did not solve for *x*, which is finding the value of a variable based on its placement into an equation; rather, in the first five chapters we learned to add, subtract, multiply, square, divide, factor and square-root various numerical and algebraic terms and evaluate them in the agreed-upon convention of mathematicians in preparation for this chapter. Now we shall solve for x and explore the mechanisms, or algorithms, by which we can so do.

The Additive Inverse Property

$a + -a = 0$ any number or term plus its additive inverse equals zero.

The Multiplicative Inverse Property

$a \cdot \dfrac{1}{a} = 1$ any number or term times its multiplicative inverse equals one.

From this it follows that

$\dfrac{a}{b} \cdot \dfrac{b}{a} = 1$ any rational number or term times its multiplicative inverse equals one.

and

$\dfrac{a}{b} \div \dfrac{c}{d} = \dfrac{a}{b} \cdot \dfrac{d}{c}$ any fraction divided by another equals the first fraction multiplied by the reciprocal of the second fraction.

In its basic and most simplistic stages, algebra can be reduced to a science by means of following a set algorithm with almost no variation. The algorithm for isolating an unknown variable on one side of an equals sign and equating it to a number is based on the idea of inverse operations: whatever operations were required to build the equation can be undone by performing the inverse of those operations. In the first four chapters, we have seen that the following pairs of operations are inverses of each other: addition and subtraction; multiplication and division; multiplication and factoring; squaring and square-rooting; cubing and cube-rooting; raising to the *nth* power and taking the *nth* root. So,

$3 + 4 = 7,$	$7 - 4 = 3$
$3 \cdot 4 = 12,$	$12/4 = 3$
$3^2 = 9,$	$\sqrt{9} = 3$

An equation is a balance scale; whatever is on one side of the equals sign has the same value, the same weight, as whatever is on the other side. When performing or undoing an operation to one side of an equation, one must perform or undo the *same thing* to the other side of the equation; otherwise, the balance is skewed and an equation no longer exists. To understand the truth and validity of this principle, consider the following examples:

$2 + 5 = 7$ Add three to both sides.
$3 + 2 + 5 = 3 + 7$
$10 = 10$

$2 + 5 = 7$ Subtract three from both sides.
$2 + 5 - 3 = 7 - 3$
$4 = 4$

$10 + 5 = 15$ Multiply every term on both sides by negative two.
$-2(10 + 5) = -2(15)$
$-2(10) + -2(5) = -2(15)$
$-20 + -10 = -30$
$-30 = -30$

$15 + 3 = 18$ Divide every term on both sides by three.
$(15 + 3)/3 = 18/3$
$(15/3) + (3/3) = (18/3)$
$5 + 1 = 6$
$6 = 6$

$\sqrt{60-11} = 7$ Square both sides of the equation.
$(\sqrt{60-11})^2 = 7^2$
$60 - 11 = 7^2$
$49 = 49$

$$3^3 + 9(6) = 81 \qquad \text{Take the square root of both sides of the equation.}$$
$$\sqrt{(27 + 54)} = \sqrt{81}$$
$$\sqrt{81} = \sqrt{81}$$
$$9 = 9$$

Anticipating forthcoming equations, the student may ask why addition and subtraction of a term affects only one term on each side of an equation, whereas multiplication and division is performed to every term on, and squaring and taking square roots is performed to the entirety of, both sides of the equation. In all cases, the answer is the implied set of parentheses around the entirety of each side of the equation. When multiplication and division are performed to both sides, the distributive property is invoked and every term on both sides of the equation is affected. When addition or subtraction is performed to both sides, the terms join an existing addition or subtraction chain on that side and only the like terms are affected. In all cases, the question at each step of the solution is whether each side of an equation has as its outer chain an addition chain or a multiplication chain and which operation is being done to either type of chain. In the following subchapters, we explore solving single variable first degree equations. As a reminder, a first degree equation is one in which the variable, x, is raised to the first power. The value of x is determined by its placement into an equation; in these sections, we will find the value of x. Bear in mind that it is only with first degree equations that x can be found by unlinking the addition or subtraction links. With second or higher degree equations, unlinking the addition/subtraction links in the chain is only the first step, done only when needed to move all terms to one side of the equation; as soon as zero remains on the other side of the equation, factoring is invoked in order to solve for the values of x. Solving higher order equations requiring the skill of factoring will follow in later subchapters.

Indeed, some of the student's former anxiety associated with solving equations may have already begun to rear its ugly head. Bear in mind that the student's ability to perform the operations of algebra has been solidly developed such that the anxiety is nothing more than false evidence appearing real. Old errors have been corrected; uncertainty and confusion have been replaced with sure knowledge. If the student has diligently worked through the first five chapters of this book, fearlessly proceed into the next sections.

Solving Single Variable First Degree Equations in One Step

In this section and the ones that follow, we explore solving single variable first degree equations. As a reminder, a first degree binomial is an equation where the variable, x, is raised to the first power. The value of x is determined by its placement into an equation; in these sections, we will find the value of x. Bear in mind that it is only with first degree equations that an addition chain can be undone by unlinking the addition or subtraction links. Solving higher order equations will follow in a later subchapter.

Ex. 1. We wish to find the value of x in the equation $x + 14 = 25$.

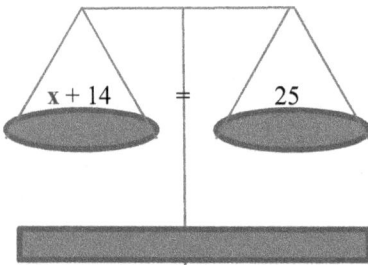

We accomplish the isolation of the variable by undoing the operations that were done to the variable, and by undoing the same thing to the other side of the equals sign. On the left, 14 was added to x; this short addition chain is undone, or unlinked, with subtraction of 14.

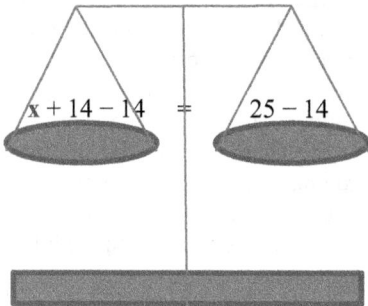

Because $14 - 14 = 0$, x remains alone on the left side. We subtract 14 from the right side, too, in order to keep the scales in balance. One way to look at this equation is to say, "I have one apple and $14; this equals $25. Therefore, what is this apple worth?" We take away $14 from both sides of the balance scale to find that the apple is worth $11.

To check if your solution is correct, substitute the value you found for x into the original equation to see if both sides are equal.

$x + 14 = 25$
$11 + 14 = 25$
$25 = 25$ ✓

Ex. 2. We wish to find the value of x in the equation

$$x - 6 = 13$$

This time, 6 is subtracted from x. This short subtraction chain is unlinked with addition of 6, from the left because $-6 + 6 = 0$, and from the right to keep the scales in balance.

$$x - 6 + 6 = 13 + 6$$
$$x = 19$$

To check if your solution is correct, substitute the value you found for x into the original equation to see if both sides are equal.

$x - 6 = 13$
$19 - 6 = 13$
$13 = 13$ ✓

Ex. 3. We wish to find the value of x in the equation

$-5x = 35$

This time, -5 is being multiplied to x. This short multiplication chain is unlinked with division by -5 to both sides of the equation: on the left side because $-5/-5 = 1$, leaving x isolated; and on the right side to keep the scales in balance.   It is critical to understand that *when unlinking multiplication with division,* ***any negative coefficient remains negative*** *because we are unlinking a multiplication chain and NOT a link in an addition or subtraction chain.*

$$\frac{-5x}{-5} = \frac{35}{-5}$$

$x = -7$

To check if your solution is correct, substitute the value you found for x into the original equation to see if both sides are equal.

$-5x = 35$
$-5(-7) = 35$
$35 = 35$ ✓

Ex. 4. We wish to find the value of x in the equation

$$\frac{x}{^-6} = 9$$

Here, x is being divided by 6. We isolate x by undoing division by – 6 with multiplication by – 6; we multiply – 6 to the other side of the equation, too. Again, *we keep the negative sign negative when we unlink – 6 because we are unlinking a division link and not an addition or subtraction link.*

$$^-6 \cdot \frac{x}{^-6} = 9(^-6), \qquad x = -54$$

To check if your solution is correct, substitute the value you found for x into the original equation to see if both sides are equal.

$$\frac{x}{^-6} = 9$$

$$\frac{-54}{-6} = 9, \qquad 9 = 9 \checkmark$$

Try these problems. Note that some answers will be integers, while other answers will be signed fractions, reducible or irreducible, or decimals.

1. $x + 8 = 24$	13. $x + 26 = 1$	25. $x + -16 = -7$
2. $x - 8 = 24$	14. $x - 26 = 1$	26. $x - ^-16 = -7$
3. $8x = 24$	15. $26x = 1$	27. $-16x = -7$
4. $\frac{x}{8} = 24$	16. $\frac{x}{26} = 1$	28. $\frac{x}{^-16} = -7$
5. $x + 6 = 12$	17. $x + 3 = -25$	29. $x + 24 = ^-8$
6. $x - 6 = 12$	18. $x - 3 = -25$	30. $x - 24 = ^-8$
7. $6x = 12$	19. $3x = -25$	31. $24x = ^-8$
8. $\frac{x}{6} = 12$	20. $\frac{x}{3} = -25$	32. $\frac{x}{24} = ^-8$
9. $x + 6 = 11$	21. $x + ^-4 = -26$	33. $x + -.9 = 2.7$
10. $x - 6 = 11$	22. $x - ^-4 = -26$	34. $x - ^-.9 = 2.7$
11. $6x = 11$	23. $^-4x = -26$	35. $-.9x = 2.7$
12. $\frac{x}{6} = 11$	24. $\frac{x}{^-4} = -26$	36. $\frac{x}{^-.9} = 2.7$

130

Solving Single Variable First Degree Equations in Two Steps

Solving an equation in two steps involves undoing the last step first and undoing the first step last. The process is much like dressing your feet: first socks go on, then shoes. When feet are undressed, first the shoes come off, then the socks. So it is with isolating a variable in a two-step equation: the variable is first multiplied to or divided by some number, creating a short, two-link multiplication chain; this monomial is then linked to some number in a two-link addition or subtraction chain. If the inverse of putting on is removing, then the inverse of addition is subtraction, the inverse of subtraction is addition, the inverse of multiplication is division and the inverse of division is multiplication. So first the addition chain is unlinked with subtraction or the subtraction chain is unlinked with addition; then the multiplication chain is undone with division or division is undone with multiplication. The same steps are done simultaneously to the other side of the equals sign in order to maintain the balance of the equation.

Ex. 1. $7x + 6 = 41$

To build the side of the equation containing x, x was first multiplied by 7 and then 6 was added to it; this follows the order of operations whereby multiplication chains are formed first and then unlike terms are linked together in an addition chain. As a reminder, performing the inverse of an operation moves the student towards isolating the variable; performing that same operation to the other side of the equals sign keeps the equation in balance.

$$7x + 6 = 41$$

To isolate x, addition of 6, the last step in building the expression, is first unlinked with subtraction of 6 from both sides.

$$\frac{-6 \quad -6}{\frac{7x}{7} = \frac{35}{7}}$$

Multiplication by 7, the first step in building the expression, is unlinked last with division by 7 to both sides.

$$x = 5$$

To check if your solution is correct, substitute the value you found for x into the original equation to see if both sides are equal.

$7x + 6 = 41$
$7(5) + 6 = 41$
$35 + 6 = 41$
$41 = 41 \checkmark$

Ex. 2.   8x – 4 = 84   Subtraction of 4 is unlinked with addition of 4.
          +4   +4
         ─────────
          8x   = 88
          ──   ──
          8   8   Multiplication by 8 is unlinked with division by 8.

          x   =   11

To check if your solution is correct, substitute the value you found for x into
the original equation to see if both sides are equal.

8x – 4 = 84
8(11) – 4 = 84
88 – 4 = 84
84 = 84 ✓

Ex. 3. $\frac{7}{8}$ x + 4 = 11
                                    Addition of 4 is unlinked with
                – 4   – 4   subtraction of 4 to both sides.
          ──────────────

          $\frac{8}{7}$ $\frac{7}{8}$ x   = 7 $\frac{8}{7}$ Multiplication of $\frac{7}{8}$ is unlinked with
                                                                 multiplication of its reciprocal, $\frac{8}{7}$,
                                                                 to both sides.
          x   =   8

To check if your solution is correct, substitute the value you found for x into
the original equation to see if both sides are equal.

$\frac{7}{8}$ x + 4 = 11
$\frac{7}{8}$(8) + 4 = 11
7 + 4 = 11
11 = 11 ✓

Ex. 4. $\frac{1}{3}$ x – 4 = 20
                                    Subtraction of 4 is unlinked with
                + 4   + 4   addition of 4 to both sides.
          ──────────────

          3• $\frac{1}{3}$ x = 24 • 3   Multiplication of $\frac{1}{3}$ is unlinked with
                                          multiplication of its reciprocal, 3,
                                          to both sides.
          x = 72

132

To check if your solution is correct, substitute the value you found for x into the original equation to see if both sides are equal.

$\frac{1}{3}x - 4 = 20$

$\frac{1}{3}(72) - 4 = 20$

$24 - 4 = 20$

$20 = 20$ ✓

1. $2x + 4 = 12$	9. $-4x + 7 = -21$	17. $14x + 8 = -3$
2. $2x - 4 = 12$	10. $-4x - 7 = -21$	18. $14x - 8 = -3$
3. $\frac{x}{2} + 4 = 12$	11. $\frac{x}{-4} + 7 = -21$	19. $\frac{x}{14} + 8 = -3$
4. $\frac{x}{2} - 4 = 12$	12. $\frac{x}{-4} - 7 = -21$	20. $\frac{x}{14} - 8 = -3$
5. $-3x + 6 = 18$	13. $22x + 5 = -16$	21. $.6x + .6 = -3.0$
6. $-3x - 6 = 18$	14. $22x - 5 = -28$	22. $.6x - .6 = -3.0$
7. $\frac{x}{-3} + 6 = 18$	15. $\frac{x}{22} + 5 = -1$	23. $\frac{x}{.6} + .6 = -3.0$
8. $\frac{x}{-3} - 6 = 18$	16. $\frac{x}{22} - 5 = -1$	24. $\frac{x}{.6} - .6 = -3.0$

Solving Single Variable First Degree Equations in Three or More Steps

Sometimes a given equation will have uncombined like terms on one or both sides of the equation. All like terms must first be combined before the procedures in the previous subsections are effected. It is critical to understand that *additive inverses are NOT performed* when uncombined like terms are *already* present on one side of the equals sign. Rather, like terms are combined in the usual manner; *after the shortest possible addition chain of unlike terms has been formed* in one or both sides, the procedures for isolating the variable by applying inverses in the previous subsections can be applied.

Ex. 1.

$$3x + 2x - 12 = 8$$
$$5x - 12 = 8$$
$$+ 12 \quad\quad + 12$$
$$5x = 20$$
$$x = 4$$

Combine like terms.
Unlink subtraction with addition; do the same thing to the other side of the equation.

Unlink multiplication with division to both sides.
The check is left to the student.

133

Ex. 2. $14x + 21 - 2x = -15$ Combine like terms; recall that the minus sign is
$\qquad 12x + 21 = -15$ attached to 2x.
$\qquad \underline{\quad -21 \quad -21}$ Unlink addition with subtraction; do the same thing
$\qquad 12x \qquad = -36$ to the other side of the equation.
$\qquad x = -3$ Unlink multiplication with division to both sides.
                                    The check is left to the student.

Realize that uncombined like terms are not restricted to variables; they may also be constants.

Ex. 3. $24 + 3x - 4 = -10$ Combine like terms.
$\qquad 20 + 3x = -10$ Unlink addition with subtraction; do the same thing
$\qquad \underline{-20 \qquad -20}$ to the other side of the equation.
$\qquad 3x = -30$ Unlink multiplication with division to both sides.
$\qquad x = -10$ The check is left to the student.

After solving the following problems, check your solutions by substituting your value for x into the original equations.

1. $9x + 13 - 3x = -11$
2. $3x - 10 + 7x = 40$
3. $3x + 6 - 5x = 20$
4. $5x + 13 - 8x = -14$
5. $3x + 1 + x + 4 = 29$
6. $x - 14 + x - 8 + 2x - 20 = 18$
7. $3x - 4 - 5x - 9 + 7x - 7 = 25$
8. $6x - 2 - 10x + 1 + 12x - 4 - x = -12$
9. $-4x + 5 - 7x + 2 - x + 8 = -14$
10. $13x - 7 + 6 - 8 + 4x + 1 = 21$

Sometimes equations have variables on both sides. To accomplish the isolation of x, all variables must be moved to one side of the equation using the principle of inverses; all constants must be moved to the other side by the same method. It is in this section in particular where the diehard habit of trying to combine apples with dollars may rear its ugly head. Remain mindful of all the hard work you have done in the first two chapters, ever recalling the concept of unlike terms linked together in an addition chain. The goal is to unlink the addition or subtraction links with the inverse operation in order to move the apples to one side of the equation and the dollars to the other side. The last step will be to unlink the multiplication or division link with its inverse to discover the dollar value of one apple.

134

Ex. 4a. $3x + 9 = 5x - 5$

      $\underline{-3x \qquad\quad -3x}$

         $9 \;\; = 2x - 5$

      $\underline{+5 \qquad\qquad +5}$

       $14 \;\; = 2x$

        $7 \;\; = \; x$

Unlink three-x (or three apples) from both sides of the equation so that no x terms remain on the left side.
All constants (or dollars) must move to the left side: unlink subtraction of five with addition of five to both sides of the equation.
Unlink multiplication by two with division by two to both sides.
The check is left to the student.

We solve the same problem, but this time the x terms will be moved to the left side and the constants will be moved to the right side, demonstrating the unimportance of which side contains the x, so long as the other side contains the constant. This illustrates the symmetric property of equality, whereby a = b is the same as b = a.

Ex. 4b. $3x + 9 = \;\; 5x - 5$

      $\underline{-5x \qquad\quad -5x}$

    $-2x + 9 \;\; = \;\; -5$

        $\underline{-9 \qquad\quad -9}$

    $-2x \;\; = \; -14$

       $x \;\; = \;\; 7$

Unlink five-x (or five apples) from both sides of sides of the equation so that no x terms remain on the right side. All constants must move to the right side: unlink addition of nine with subtraction of nine to both to both sides. Unlink multiplication by negative two with division by negative two to both sides. The check is left to the student.

For students who are confident in distinguishing apples from dollars, two steps may be done in one.

Ex. 5. $\underline{\text{constants this way}}$ →

        ← $\underline{\text{variables that way}}$

        $2x + 14 \;\; = \; -7x + 5$

        $\underline{+7x - 14 \;\; +7x - 14}$

        $9x \qquad\; = \; -9$

          $x \qquad = \; -1$

Unlink $-7x$ with $+7x$ to both sides; unlink 14 with -14 to both sides.

Unlink multiplication by 9 with division by 9 to both sides.
The check is left to the student.

11. $5x = 3x + 8$

12. $-7x = -3x + 12$

13. $-4x = 12x - 8$

14. $11x + 9 = 17x - 9$

15. $-9x - 6 = -7x - 24$

16. $-5x - 18 = -3x - 7$

17. $18x - 81 = -18x - 9$

18. $\dfrac{3x - 10}{5} = \dfrac{-x - 11}{5}$

19. $\dfrac{3x}{4} - 18 = 11 - \dfrac{2x}{7}$

20. $\dfrac{-2x - 3}{3} = 9 - \dfrac{2x}{5}$

Some problems require use of the distributive property on one or both sides of the equation. Recall from Chapter Three that the distributive property is nothing more than multiplication. The four-step procedure is: (i) use distributive property – **MULTIPLY**; (ii) combine like terms – **ADD**; (iii) **Unlink ADDITION OR SUBTRACTION** with its inverse to both sides, moving variables to one side and constants to the other side; and (iv) **Unlink MULTIPLICATION OR DIVISION** with its inverse to both sides. The first two steps require working each side of the equation separately to express the shortest possible addition chain on both sides. In contrast, the last two steps require performing the same operation simultaneously to both sides.

Ex. 6.

$$4(x - 7) + 24 = 2(x + 3)$$
$$4x - 28 + 24 = 2x + 6$$

Remove parentheses with the distributive property - **MULTIPLY**. Combine like terms – **ADD**.

$$4x - 4 = 2x + 6$$
$$\underline{-2x + 4 = -2x + 4}$$

Unlink ADDITION of $+ 2x$ on the right with $-2x$ to both sides; unlink -4 on the left with $+4$ to both sides.

$$2x = 10$$
$$x = 5$$

Unlink MULTIPLICATION of 2 with division of 2 to both sides.

Ex. 7.

$$-4(x - 7) + 6x = -3(x + 3)$$

$$-4x + 28 + 6x = -3x - 9$$

Remove parentheses with the distributive property – **MULTIPLY**. Combine like terms – **ADD**.

$$2x + 28 = -3x - 9$$
$$\underline{+3x - 28 = +3x - 28}$$
$$5x = -37$$
$$x = -7.4 \text{ or } -7\,^2/_5$$

Unlink ADDITION of $-3x$ on the right with $+3x$ to both sides; unlink $+28$ on left with -28 to both sides. **Unlink MULTIPLICATION** of 5 with division of 5 to both sides.

When solving the problems below, remember that the minus sign is ALWAYS attached to the term that follows it; when a minus sign is followed by a term in front of parentheses, the term should be distributed as a negative to the terms inside parentheses; the terms inside parentheses following minus signs should likewise be treated as negatives.

21. $3(x + 5) = -30$
22. $2(4x - 7) = 4x + 14$
23. $8(x + 1) = -4(x - 14)$
24. $5(x - 30) = -3(x - 6)$
25. $-9(x + \frac{1}{2}) - 7x = 4(x - 4) + 5$

136

26. $8(x-2)+5x+12=12(x+\frac{2}{3})-7x-4$
27. $3(x-6)-6(x-14)+6=12(x-6)-5(x-7)-3x$
28. $2(2x+80)+3(3x+30)+25=2(2x-105)-4(4x+30)-20$

Solving Single Variable First Degree Inequalities in One or More Steps

Solving inequalities involves the same procedures as solving equations, with one exception: **when both sides of an inequality are multiplied or divided by any negative term, the inequality symbol is reversed.**

Ex. 1. $3<5$

Numbers moving from left to right on the number line increase in order. Now multiply both sides of the inequality by -1:

$-3>-5$ or $-5<-3$

Ex. 2. $3x+9 \geq 5x-5$

Note that unlinking negative terms by the additive inverse property does not reverse the inequality symbol.

$$\underline{-5x\ -9\ \ -5x\ -9}$$

Note that unlinking a negative term by the multiplicative inverse property *does* reverse the inequality symbol.

$-2x\ \ \geq -14$

$x\ \ \leq\ 7$

The solution to an inequality, unlike an equation, is graphed as a ray on a number line.

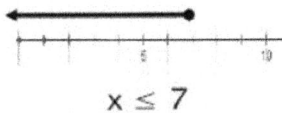

$x \leq 7$

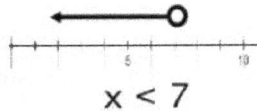

$x < 7$

The endpoint of the ray is closed when the symbols $\leq$ or $\geq$ are used, indicating that the number together with all points covered by the arrow is part of the solution set. If the endpoint had been open, with the symbols $<$ or $>$ used, the number would not have been part of the solution set.

1. $x+12>8$
2. $x-12<8$
3. $12x \geq 8$
4. $\dfrac{x}{12} \leq 8$

5. $x+3.1<-.6$
6. $x-3.1>-.6$
7. $3.1x \leq -.6$
8. $\dfrac{x}{3.1} \geq -.6$

9. $x+-2.5 \geq 1.0$
10. $x- {}^-2.5 \leq 1.0$
11. $-2.5x < 1.0$
12. $\dfrac{x}{{}^-2.5} > 1.0$

137

13. $-9x + 4 \geq 5$

14. $-9x - 4 \leq 5$

15. $\dfrac{x + 4}{-9} < 5$

16. $\dfrac{x - 4}{-9} > 5$

17. $1.5x + .5 > 2.5$

18. $1.5x - .5 \geq 2.5$

19. $\dfrac{x + .5}{1.5} < 2.5$

20. $\dfrac{x - .5}{1.5} \leq 2.5$

21. $-.8x + .6 < 7.0$

22. $-.8x - .6 \leq 7.0$

23. $\dfrac{x + .6}{^-.8} \geq 7.0$

24. $\dfrac{x - .6}{^-.8} > 7.0$

25. $11(x + 1) + x \leq -8(x + ½)$

26. $24(x + ½) - 5x \geq 8(x + 2) + 5x$

Solving Single Variable Second Degree Equations

In second degree equations, there are not one but rather two values of x which will make the equation true when those values are substituted, one at a time, into the equation. Unlinking the addition links in these and higher degree equations is never a final step but only an initial step, and only when necessary to move all terms of the equation to one side of the equals sign. As soon as all terms are moved to one side of the equation, factoring is needed in order to find both values of x.

Ex. 1. The equation

$$x^2 = -3x - 2$$

is rewritten by unlinking $-3x$ with its additive inverse to both sides and unlinking -2 with its additive inverse to both sides.

$$x^2 + 3x + 2 = 0$$

Once an addition chain comprising a quadratic trinomial is formed, we factor; this becomes
$$(x + 1)(x + 2) = 0$$

Assume for a moment that $a = (x + 1)$ and $b = (x + 2)$. We now have a simplified equation

$$ab = 0$$

Note that this equation equals zero when either $a = 0$, $b = 0$ or both equal zero. This is the principle behind setting each factor equal to zero, and so we do now:

$$x + 1 = 0 \qquad\qquad x + 2 = 0$$

We solve each of these simple linear equations by unlinking the plus signs with the additive inverse to both sides.

$$x = -1 \text{ and } x = -2$$

By substitution,

$x^2 + 3x + 2 = 0$,
$(-1)^2 + 3(-1) + 2 = 0$
$1 - 3 + 2 = 0$
$0 = 0$

$x^2 + 3x + 2 = 0$
$(-2)^2 + 3(-2) + 2 = 0$
$4 - 6 + 2 = 0$
$0 = 0$

There is, in fact, an exception to the rule that all quadratic equations must be solved by moving all terms to one side of the equation. That exception arises when the quadratic equation does not contain a first degree monomial, in other words, a term with an x to the invisible first power. In that instance, addition or subtraction is unlinked to move the additive inverse of the constant to the side containing zero. Then the square root of both sides is taken to find the two values of x.

Ex. 2.　　$4x^2 - 9 = 0$ 　　　Unlink -9 by adding it to both sides. Unlink 4
　　　　　　$4x^2 = 9$ 　　　　　by dividing both sides by it.
　　　　　　$x^2 = {}^9/_4$ 　　　　Take the square root of both sides.
　　　　　　$x = \pm\, {}^3/_2$

The plus sign above the minus sign is a common abbreviation in algebra used to denote that both $+\, {}^3/_2$ and $-\, {}^3/_2$ are the correct values of x (recall that negative times negative equals positive). A check is always in order.

$4x^2 - 9 = 0$
$4\left({}^3/_2\right)^2 - 9 = 0$
$4\left({}^9/_4\right) - 9 = 0$
$9 - 9 = 0$
$0 = 0$

$4x^2 - 9 = 0$
$4\left({}^{-3}/_2\right)^2 - 9 = 0$
$4\left({}^9/_4\right) - 9 = 0$
$9 - 9 = 0$
$0 = 0$

Note in the next example that this exception does not hold when the constant term is missing. We factor, not by using FOIL in reverse, but by finding the greatest common factor using the distributive property in reverse; then we set each factor equal to zero to solve for the values of x.

$x^2 = 5x$ 　　　　　Unlink 5x with its additive inverse to both
$x^2 - 5x = 0$ 　　　sides.

139

$x(x-5)=0$ Factor x from both terms on the left.

$x = 0 \mid x - 5 = 0$ Set each factor equal to zero.

$ x = 5$

1. $2x^2 = x^2 - 11x - 18$
2. $3x^2 - 9x = 2x^2 - 8$
3. $3x^2 - 6x = 2x^2 + 27$
4. $5x^2 = 4x^2 + 12x - 35$
5. $2x^2 = x + 15$
6. $2x^2 - 7x - 10 = -3x^2 + 6x + 18$
7. $4x^2 + 5x = x^2 + 2$
8. $2x^2 - 7x = 18 + 8x - x^2$

Word Problems

Addition Phrases	Subtraction Phrases
more than added to increased by the sum of combined together total of plus	* less than, fewer than * subtracted from decreased by the difference between less minus
Multiplication Phrases	**Division Phrases**
twice the product of times of percent of factors the square of (repeated multiplication)	half of the quotient of divided by per out of percent (divide by 100)

Most students find word problems challenging to solve. The reasons for this are that specific phrases indicate a specific operation to be used, several such phrases appear in any given problem, and students have trouble distinguishing one operation from another. The format of this text was designed to remedy this issue. Once the essence of an operation is understood, it simply becomes a matter of memorizing which phrase goes

140

with which operation and performing a literal, word-for-word translation from English into algebra.

By now you realize that, on either one or both sides of an equals sign, most equations are nothing more than an addition/subtraction chain connecting multiplication chains. The construction of your equations from words should be no different. The key is to identify the terms which make up the multiplication chains and to determine the positions of equals sign and the addition or subtraction links. This can often be accomplished by performing a literal, word-for-word translation from English into algebra.

Ex.1. Five more than three times a number is 35.
Solution: The number is unknown; assign x to the number.

Five more than three times a number is 35.   Diagram the sentence.

$5 + 3x = 35$ Translate from left to right, number for number and
$3x = 30$ phrase for operation. "Is" means "equals" (=).
$x = 10$

Ex. 2. When a number is increased by 20, the result is 80 decreased by the number.

Solution: Assign x to the unknown number and diagram the sentence.

Let x = the number

When a number is increased by 20, the result is 80 decreased by the number.

$x + 20 = 80 - x$ Translate from left to right, number for number and phrase for
                    operation. "Is" means "equals" (=).
$2x = 60$ Solve the equation.
$x = 30$ The check is left to the student.

Ex. 3. One number is four more than another number. The square of the smaller is 56 less than the square of the larger. Find the numbers.

The strategy for solving word problems where several unknowns are expressed in terms of x is *to assign x to the second unknown appearing in the problem*. The first unknown is being compared to the second, and a third unknown, if present in the problem, will be compared to the first or second,

141

so all other unknowns assume a variable expression in terms of the second unknown, which gets the x.

Let   x   = the smaller number
            $x + 4$ = the larger number

In English, the phrases *less than* or *subtracted from* express the minuend and subtrahend in reverse order. You must rewrite the order: move the minuend in front of the subtraction symbol and move the subtrahend after it.

subtrahend minus sign   minuend

56 less than the square of the larger

$$(x)^2 = (x + 4)^2 - 56$$

$x^2 = x^2 + 8x + 16 - 56$	Subtract x^2 from both sides; combine like terms.
$0 = 8x - 40$	Unlink -40 with its additive inverse to both sides.
$40 = 8x$	Unlink 8 with its multiplicative inverse to both sides.
$x = 5$	

When the word problem results in a quadratic equation, two solutions may be possible. Both must be checked. If one solution does not check, it is rejected as an *extraneous solution*.

Ex. 4. The larger of two numbers is five more than twice the smaller. Twice the square of the smaller is eight times the larger. Find the numbers.

Solution: x = smaller
                $2x + 5$ = larger

$2x^2 = 8(2x + 5)$	Distribute (multiply) 8 to the addition chain.
$2x^2 = 16x + 40$	Unlink $16x$ and 40 with their additive inverses.
$2x^2 - 16x - 40 = 0$	When all terms in a quadratic addition chain have a
$2(x^2 - 8x - 20) = 0$	common factor, it must be factored from all terms
$2(x - 10)(x + 2) = 0$	in the equation in order for FOIL factoring to work.
$2 \neq 0, \; x = 10, \; x = -2$	Factor using FOIL in reverse; set each factor $= 0$.
Check: $2(10)^2 = 8[2(10) + 5]$	Note that two never equals zero and is rejected.
$200 = 200 \checkmark$	In this problem, both solutions check. In problem 6
$2(-2)^2 = 8[2(-2) + 5]$	below, there is an extraneous solution; check for it.
$8 = 8 \checkmark$	

142

1. If seven times a number is increased by two, the result is 13 less than 10 times the number. Find the number.
2. The larger of two numbers is 7 more than the smaller. Four times the larger is thirty-six more than three times the smaller. Find the numbers.
3. The larger of two numbers is seven less than twice the smaller. Seven times the larger is four less than nine times the smaller. Find the numbers.
4. The larger of two numbers is five more than the smaller. Six times the larger decreased by eight is eight times the smaller. Find the numbers.
5. One number is seven more than five times another number. Together they add up to fifty-five. Find the numbers.
6. The larger of two numbers is six more than the smaller. Four times the square of the smaller is 24 decreased by the larger. Find the numbers.

Consecutive Integer Problems

Consecutive integer problems are among the easiest problem to solve because the assignment of the variables is always the same. The legend, or key, for consecutive integers always assigns x to the first integer, x + 1 to the second consecutive integer, x + 2 to the third consecutive integer and, if necessary, x + 3 to the fourth consecutive integer.

Whole numbers (n) – the counting numbers (1, 2, 3, 4, ..., ∞) and zero.

Integers (n) – the set of whole numbers and their negative opposites.

Consecutive integers (n) – a set of integers one immediately following the next, i.e., 2, 3 and 4 are three consecutive integers.

Legend (n) – The key used to assign variables and expressions containing variables to the unknown values in a word problem.

Ex. 1. Find three consecutive positive integers such that two times the first plus four times the second equals four more than five times the third.

Solution: First we set up the legend:

Let x   = the 1st consecutive positive integer
x + 1 = the 2nd consecutive positive integer
x + 2 = the 3rd consecutive positive integer

Next we diagram the sentence.

Find three consecutive positive integers such that two times the first plus four times the second **equals** four more than five times the third.

Using parentheses around the variable expressions is key to expressing multiplication and sets up any necessary distributive property application.

$2(x) + 4(x + 1) = 4 + 5(x + 2)$	Use the distributive property.
$2x + 4x + 4 = 4 + 5x + 10$	Combine like terms on both sides.
$6x + 4 = 5x + 14$	Unlink addition links: move the opposite of the first degree variable on the right to the left; move the opposite of the constant on the left to the right.
$x = 10$	Evaluate all variable expressions.
$x + 1 = 11$	
$x + 2 = 12$	

Substitute the values into the original equation to check that your solution is correct.

$2(10) + 4(11) = 4 + 5(12)$
$20 + 44 = 4 + 60$
$64 = 64$ ✓

Ex. 2. Find three consecutive positive integers such that the first times the second equals sixty-seven less than the square of the third.

Solution: Establish the legend

Let x = the 1st consecutive positive integer
$x + 1$ = the 2nd consecutive positive integer
$x + 2$ = the 3rd consecutive positive integer

Next diagram the sentence.

Find three consecutive positive integers such that the first times the second **equals** sixty-seven less than the square of the third.

In English, the phrases *less than* or *subtracted from* express the minuend and subtrahend in reverse order; you must correct the order by moving the minuend in front of the subtraction symbol and moving the subtrahend after it.

subtrahend   minus sign   minuend

sixty-seven less than the square of the third

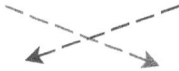

$(x)(x + 1) = (x + 2)^2 - 67$ Use the distributive property.

$x^2 + x = x^2 + 4x + 4 - 67$ Combine like terms on both sides.

$x^2 + x = x^2 + 4x - 63$ Unlink addition links: remove the x-squared terms from both sides; move the opposite of the first degree variable on the right to the left.

$+ x = 4x - 63$

$\underline{- 4x \quad - 4x}$

$- 3x = - 63$ Divide both sides by negative three.

$x = 21$ Evaluate all variable expressions.

$x + 1 = 22$

$x + 2 = 23$

Substitute the values into the original equation to check that your solution is correct.

$(x)(x + 1) = (x + 2)^2 - 67$

$(21)(22) = (23)^2 - 67$

$462 = 629 - 67$

$462 = 462$

1.   Find three consecutive positive integers such that the first times the second equals 43 less than the square of the third.

2.   Find three consecutive positive integers such that four times the first plus three times the second equals five more than five times the third.

3.   Prove that, given any three consecutive integers, the first times the third is always one less than the square of the second.

4. Prove that, given any four consecutive integers, the first times the fourth is always two less than the second times the third.

Problems 3 and 4 result in an *identity*, an equation true for all values of *x*.

145

Motion Problems

Motion problems are presented as a distinct subchapter because they give students an unusually difficult time; yet, they present an excellent example of how word problems translate into the formation of multiplication chains within an addition chain.

In all motion problems, the formula $d = r\,t$ is used, where d represents distance traveled, r represents the rate of speed and t represents the time traveled. Note how the product, distance, is a *different term* from the original two factors, rate of speed times time. The underlying multiplication chain to be formed is the substitution of rt for d, since information on distances traveled by the different moving objects is often provided verbally in terms of speed and time. Motion problems in this text are categorized by two types; regardless of the type, the distance traveled by each object is always to be found in the *check*.

Type I. Time, the unknown, is EQUAL for both moving objects. Distance and rate of speed are unequal.

      A. Two moving objects start from the same point moving in opposite directions. Find the time at which they are some given (constant) distance apart.

$$r_1 t \qquad\qquad\qquad r_2 t$$

$$\underbrace{r_1 t}_{\text{distance object 1}} \;+\; \underbrace{r_2 t}_{\text{distance object 2}} \;=\; \underbrace{k\,(\text{constant})}_{\text{given total distance}}$$

Again, the rates of speed are unequal, and t is the unknown variable for both moving objects.

Ex. 1. Two cars leave from the same point at the same time moving in opposite directions. The first car travels at a constant rate of 56 mph. The second car travels at a constant rate of 60 mph. At what time will the cars be 464 miles apart?

Solution:

$56t + 60t = 464$	Combine like terms.
$116t = 464$	Unlink 116 with its multiplicative inverse.
$t = 4$ hrs.	

Check: 56(4) + 60(4) = 464
224 mi. + 240 mi. = 464 mi.   Note that the distance traveled by each
        464 = 464   object is found in the check.

  B. Two moving objects start a given distance apart, move toward
      each other and either (i) close the distance completely, that is,
      pass each other, or (ii) are still some given distance apart at time
      t.

  (i) r_1t r_2t

$$\underset{\text{distance object 1}}{\underline{r_1t}} \quad + \quad \underset{\text{distance object 2}}{\underline{r_2t}} \quad = \quad \underset{\text{given total distance}}{\underline{k \text{ (constant)}}}$$

  (ii) r_1t r_2t

$$\underset{\text{distance object 1}}{\underline{r_1t}} + \underset{\text{distance object 2}}{\underline{r_2t}} + \underset{\substack{\text{distance apart} \\ \text{(constant)}}}{\underline{c}} = \underset{\text{given total distance}}{\underline{k \text{ (constant)}}}$$

Ex. 2. A car leaves Florence for Venice traveling 24 kph. At the same
         time, a car leaves Venice for Florence traveling 27 kph. If the distance
         between Venice and Florence is 255 km, at what time will the cars pass
         each other?

Solution: $24t + 27t = 255$ Combine like terms.
                $51t = 255$ Unlink 51 with its multiplicative inverse.
                $t = 5$ hrs.

Check: $24(5) + 27(5) = 255$
              120 km + 135 km = 255 km
              255 = 255

Ex. 3. A car leaves Brussels for Paris traveling 45 kph. At the same time, a
         car leaves Paris for Brussels traveling 31 kph. If the distance between
         Paris and Brussels is 261 km, at what time will the cars still be 33 km
         apart?

147

Solution: $45t + 31t + 33 = 261$ Combine like terms.
                 $76t + 33 = 261$ Unlink 33 with its additive inverse.
                 $76t = 228$ Unlink 76 with its multiplicative inverse.
                 $t = 3$ hrs.

Check: $45(3) + 31(3) + 33 = 261$
         135 km $+ 93$ km $+ 33$ km $= 261$ km
                              $261 = 261$

    C. Two objects leave from the same place at the same time going in the same direction; one is faster, the other is slower. At time t, the faster object has traveled a given distance farther.

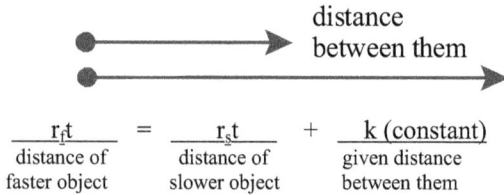

$$\underbrace{r_f t}_{\substack{\text{distance of}\\\text{faster object}}} = \underbrace{r_s t}_{\substack{\text{distance of}\\\text{slower object}}} + \underbrace{k \text{ (constant)}}_{\substack{\text{given distance}\\\text{between them}}}$$

Ex. 4.   Two cars leave from the same place at the same time traveling in the same direction. One car travels at 30 mph; the other car travels at 25 mph. At what time will the distance between them be 30 miles?

Solution: $30t = 25t + 30$ Unlink 25t with its additive inverse to both sides.
             $5t = 30$ Unlink 5 with its multiplicative inverse to both sides.
             $t = 6$ hrs.   The check is left to the student.

Type II: Distance is EQUAL for both objects. Time and rate of speed are unequal.

    A. The "overtake" problem. The slower object moves first; the faster object leaves later but catches up and overtakes the slower object. The strategy is to assign t to the second faster object and $t + t_k$ to the first slower object.

$$\underbrace{r_f t}_{\substack{\text{distance of}\\\text{faster object}}} = \underbrace{r_s(t + t_k)}_{\substack{\text{distance of}\\\text{slower object}}} \qquad \begin{array}{l}t_k = \text{extra } \textit{time} \text{ first slower object}\\ \text{started before the second}\\ \text{faster object; it represents a}\\ \text{given constant time.}\end{array}$$

148

Ex. 5.   A train leaves Chicago traveling at 60 mph. One hour later, another train leaves the same station traveling at 90 mph. In how many hours (or – a slightly longer problem – after how many miles) does the second train overtake the first?

Solution: $90t = 60 (t + 1 \text{ hr})$ Distribute 60 to the addition chain.

                   $90t = 60t + 60$ Unlink 60t with its additive inverse.

                   $30t = 60$ Unlink 30 with its multiplicative inverse.

                   $t = 2 \text{ hrs.}$

Check: $90(2) = 60(2 + 1)$ Note that the distance traveled by each object

                   $180 \text{ mi} = 180 \text{ mi.}$ continues to be found in the check.

      B.   The "round-trip" problem. One object makes a round trip at different rates of speed and in different times; distance is EQUAL. The total time is a given constant; the time traveled in each direction is represented by t and $k - t$.

Ex. 6.   A train leaves Rome for Bologna traveling at 96 kph. On its return trip it travels at 64 mph. Total travel time is 10 hours. How long did each leg of the trip take? What is the distance from Rome to Bologna?

Solution: $96t = 64 (10 - t)$ Distribute 64 to the subtraction chain.

                   $96t = 640 - 64t$ Unlink – 64t with its additive inverse.

                   $160t = 640$ Unlink 160 with its multiplicative inverse.

                   $t = 4 \text{hrs.}$

                   $10 - t = 6 \text{ hrs.}$

Check: $96(4) = 64(6)$

                   $384 \text{ km} = 384 \text{ km}$

1. Two ships leave from the same point at the same time moving in opposite directions. The first ship travels at a constant rate of 60 mph. The second ship travels at a constant rate of 70 mph. At what time will the ships be 650 miles apart?

2. A plane leaves Hong Kong for Shanghai traveling 290 kph. At the same time, a plane leaves Shanghai for Hong Kong traveling 322 kph. If the distance between Shanghai and Hong Kong is 1,224 km, at what time will the planes pass each other?

3. A plane leaves New York for Los Angeles traveling 238 mph. At the same time, a plane leaves Los Angeles for New York traveling 309 mph. If the distance between New York and Los Angeles is 2,448 miles, at what time will the planes still be 260 miles apart?

4. Two cars leave from Miami at the same time traveling in the same direction. One car travels 45 mph; the other car travels 55 mph. At what time will the distance between them be 50 miles?

5. A train leaves Barcelona traveling 80 kph. Two hours later, another train leaves the same station traveling 140 kph. How long does it take for the second train overtake the first?

6. A train leaves Bologna for Naples traveling 100 kph. On its return trip it travels 75 kph. Total travel time is 14 hours. How long did each leg of the trip take? What is the distance from Bologna to Naples?

Mixture Problems

Mixture problems provide another excellent demonstration for how multiplication changes the nature of terms being multiplied. The formation of multiplication chains linked together to form an addition chain makes up the backbone of the equation involving mixture problems.

Ex. 1. A coffee distributor wants to mix coffee which sells for $7.99/lb. with coffee which sells for $10.99/lb. to make 30 pounds of coffee to sell at $9.99/lb. How many pounds of each type of coffee should she use?

Solution: Let $\quad$ x = # of lbs. of $7.99/lb. coffee
$\quad\quad\quad\quad$ 30 − x = # of lbs. of $10.99/lb. coffee

Note that x + 30 − x = 30 lbs. which should be the total weight of our mixture. The structure of each monomial is

$$\frac{\$}{lb.} \cdot \frac{\# \; of \; lbs}{1} = \$$$

799x + 1099(30 − x) = 999(30) $\quad\quad$ Distribute 1099 to the binomial.
Cost of $\quad$ Cost of $\quad\quad$ Cost of
1st type $\quad$ 2nd type $\quad\quad$ final blend

799x + 32,970 − 1099x = 29,970 $\quad\quad$ Combine like terms.

150

$-300x + 32,970 = 29,970$ Subtract 32,970 from both sides.

$-300x = -3,000$ Divide both sides by - 300.

$x = 10$ lbs.   Find the other unknown.

$30 - x = 30 - 10 = 20$ lbs.   Check by substitution.

Check: $7.99/lb.(10 lbs.) + $10.99/lb.(20 lbs.) = $9.99(30 lbs.)

$79.90 + $219.18 = $299.70

$ 299.70 = $299.70

Ans: The distributor should use 10 lbs. of the coffee selling at $7.99/lb. and 20 lbs. of the coffee selling at $10.99/lb.

Ex. 2. How many pounds of peanuts selling at $3.49/lb. must be mixed with 40 lbs. of pistachios selling at $16.99/lb. and 40 lbs. of walnuts selling at $9.99/lb. to produce a mixture which can sell for $10.99/lb.?

Solution: Let x = # of lbs. of peanuts selling at $3.49/lb.

$349x + 1699(40) + 999(40) = 1099(80 + x)$

$349x + 67960 + 39960 = 87920 + 1099x$

$349x + 107920 = 87920 + 1099x$

$-349x \quad -87920 \quad -87920 \quad -349x$

$20000 = 750x$

$x = 26.667$ lbs.   (To obtain a precise calculation to the penny, use a decimal beyond its place value).

Check: $3.49(26.667) + 16.99(40) + 9.99(40) = 10.99(80 + 26.667)

$93.07 + $679.60 + $399.60 = $1,172.27

$1,172.27 = $1,172.27

1. A landscaper wants to mix Bermuda grass seed worth $4.00/lb. with Kentucky Bluegrass seed worth $2.00/lb for a total of 50 lbs. worth $2.50/lb. How many pounds of each should he use?

2. A tea distributor wants to mix 50 lbs. of Oolong tea selling at $23.99/lb. with Darjeeling tea selling at $28.99/lb. for a mixture that will sell at $24.99/lb. How many pounds of mixture will she get?

3. A grocer wants to make a pasta spinach salad using 10 lbs. pasta costing him $0.49/lb. and frozen spinach costing $0.99/lb. If his cost for these ingredients is to be $0.79/lb., how many pounds of spinach must he use?

The Pythagorean Theorem

A unique relationship exists among the legs and hypotenuse of every right triangle.

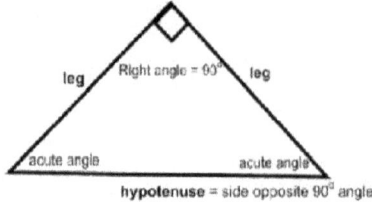

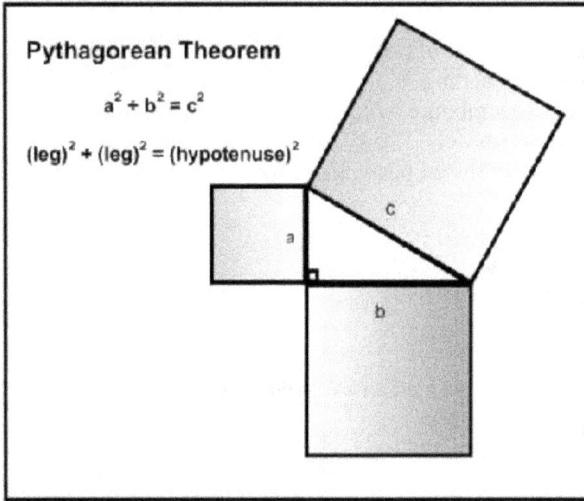

leg

Right angle = 90°

leg

acute angle

acute angle

hypotenuse = side opposite 90° angle

Pythagorean Theorem

$$a^2 + b^2 = c^2$$

$$(\text{leg})^2 + (\text{leg})^2 = (\text{hypotenuse})^2$$

a

b

c

The hypotenuse, by definition, is the side opposite the 90 degree angle of every triangle; it is the longest side of every right triangle and is always represented by the variable c in the Pythagorean Theorem formula. The shorter two sides of the right triangle are called legs. It never matters which leg is represented by a or b; it only matters that each leg is assigned to those variables as they are placed in the formula.

The concept of leg multiplied to itself, added to second leg multiplied by itself and equaling the hypotenuse multiplied to itself makes sense only when the nature of repeated multiplication and its corresponding geometric representation, namely, the square, is understood. Geometrically, the formula signifies that the sum of the areas of the squares formed by repeated multiplication of legs a and b equals the area of the square formed by repeated multiplication of the hypotenuse.

152

The truth of the Pythagorean Theorem is best illustrated in examples. Consider a right triangle with legs of 3 units and 4 units in length and a hypotenuse of 5 units in length. Then

$$a^2 + b^2 = c^2$$
$$3^2 + 4^2 = 5^2$$
$$9 + 16 = 25$$
$$25 = 25$$

Geometrically , this states that a square with area of 9 square units added to a square with area of 16 square units equals a square with area of 25 square units.

The Pythagorean Theorem is used to find the length of the side of a right triangle when the lengths of the other two sides are given. The following example shows when the length of the hypotenuse is unknown. Consider a right triangle with legs of lengths 2 and 5 respectively. By the Pythagorean Theorem,

$a^2 + b^2 = c^2$	Substitute 2 and 5 for a and b, respectively.
$2^2 + 5^2 = c^2$	Square both terms.
$4 + 25 = c^2$	Combine like terms.
$29 = c^2$	Take the square root of both sides of the equation.
$c = \sqrt{29} \approx 5.385$	

It is best to leave an irrational value in simplest radical form, unless directed otherwise.

Now consider a right triangle with hypotenuse of length 13 and leg of length 5. The length of the other leg can be found:

$a^2 + b^2 = c^2$	Substitute 5 and 13 for a and c, respectively.
$5^2 + b^2 = 13^2$	Square both terms.
$25 + b^2 = 169$	Unlink addition of 25 with subtraction of 25 to both sides.
$b^2 = 169 - 25 = 144$	Take the square root of both sides of the equation.
$b = \sqrt{144} = 12$	

Whether you are trying to find the hypotenuse or one of the legs, the last step of every Pythagorean Theorem problem is to take the square root of both sides of the equation. Try some now.

1.	$a = 30$, $c = 50$	7.	$a = 3$, $b = 7$
2.	$a = 6$, $b = 8$	8.	$a = 3$, $c = 7$
3.	$b = 40$, $c = 41$	9.	$b = 5$, $c = 9$
4.	$a = 7$, $b = 24$	10.	$b = \sqrt{5}$, $c = \sqrt{21}$
5.	$a = 8$, $c = 17$	11.	$a = \sqrt{7}$, $b = \sqrt{13}$
6.	$a = 5$, $b = 3$	12.	$a = \sqrt{8}$, $c = \sqrt{28}$

13. A baseball diamond is in fact a square. The distance from one base to the next is 90 feet. What is the linear distance from second base to home plate? Express your answer in simplest radical form; then evaluate and round to the nearest foot.

14. A 26-foot ladder leans against a building. The foot of the ladder is ten feet from the base of the building. How high up the wall does the ladder reach?

Ratio and Proportion

Definition of Ratio

A ratio is a comparison of two numbers $\frac{a}{b}$, where $b \neq 0$.

A ratio can also be expressed as

$a:b$ $\qquad$ a to b $\qquad$ a out of b

Definition of Proportion

A proportion is an equation setting one ratio equal to another.

$$\frac{a}{b} = \frac{c}{d}$$

A proportion may also be expressed as

$a:b::c:d$

$\text{extreme}_1 : \text{mean}_1 :: \text{mean}_2 : \text{extreme}_2$

a and d are called the extremes, or outer values, of the proportion; b and c are called the means, or middle values, of the proportion.

It is critical to understand that division by zero is a mathematical impossibility which cannot be accomplished to obtain an exact number but only can be interpreted to mean positive or negative infinity, which is not a number but rather a symbol representing the concept that the end is never reached because there is no end. This idea is explored in higher level mathematics courses. For purposes of this text, it is important to understand that no denominator in either a ratio or a proportion may equal zero.

Ex. 1. Consider the obvious truth of following equation

$$\frac{6}{2} = 3$$

By the second usage of the multiplicative identity property, let us divide three by 1; it still equals itself and the truth of the equation remains intact.

Extreme₁ $\frac{6}{2} = \frac{3}{1}$ Mean₂
Mean₁   Extreme₂

We can see that a proportion is nothing more than an equation simultaneously demonstrating the relationship between multiplication and its inverse operation, division. If we cross-multiply, we obtain the still-true but reconfigured equation

$$(6)(1) = (2)(3)$$

or

$$(\text{Mean}_1)(\text{Mean}_2) = (\text{Extreme}_1)(\text{Extreme}_2)$$

In words, the product of the means equals the product of the extremes.

From this it follows that:

(1) You can switch the means and the equation will still be true, provided that neither numerator or denominator in the proportion equals zero;

$$\frac{6}{3} = \frac{2}{1}$$

(2) You can switch the extremes and the equation will still be true provided that neither numerator or denominator in the proportion equals zero;

$$\frac{1}{3} = \frac{2}{6}$$

(3) You can flip the entire equation upside-down and the equation will still be true, provided that neither numerator or denominator in the proportion equals zero.

$$\frac{3}{1} = \frac{6}{2}$$

155

It is interesting to note that a proportion has a diaphanous quality to it; its truth is apparent from any perspective or position from which it is viewed, whether from right to left, left to right, top to bottom or upside down. With this in mind, a very important and time-saving shortcut follows.

Extreme = $\dfrac{(\text{Mean}_1)(\text{Mean}_2)}{\text{Other Extreme}}$ $\qquad\qquad$ Mean = $\dfrac{(\text{Extreme}_1)(\text{Extreme}_2)}{\text{Other Mean}}$

$3 = \dfrac{(1)(6)}{2}$ $\qquad$ $2 = \dfrac{(1)(6)}{3}$ $\qquad$ $6 = \dfrac{(2)(3)}{1}$ $\qquad$ $1 = \dfrac{(2)(3)}{6}$

This idea is an invaluable shortcut when solving for x; its applications extend beyond elementary algebra to basic trigonometry, differential calculus and more.

When solving for x, recall that a set of parentheses protects each numerator and each denominator of the fractions in the proportion. Thus, the distributive property or FOIL must be used. The next problem illustrates this, as well as an interesting outcome when two solutions for x result.

Ex. 2.

$\dfrac{5}{x} = \dfrac{(x+22)}{48}$ $\qquad$ Cross-multiply: use distributive property
$\qquad\qquad\qquad\qquad$ to multiply the means.

$x^2 + 22x = 240$ $\qquad$ Unlink 240 from the right with the additive
$\qquad\qquad\qquad\qquad$ inverse to both sides.
$x^2 + 22x - 240 = 0$ $\qquad$ Factor using FOIL in reverse.

$(x - 8)(x + 30) = 0$ $\qquad$ Set each factor equal to zero; solve.

$x - 8 = 0 \mid x + 30 = 0$

$\quad x = 8 \quad \mid x = -30$ $\qquad$ To check, substitute each solution into the original equation.

Check: Here is an interesting outcome: the additive inverse of each Solution equals the other variable expression in the other check.

$\dfrac{5}{8} = \dfrac{(8+22)}{48}$ $\qquad\qquad\qquad$ $\dfrac{5}{-30} = \dfrac{(-30+22)}{48}$

$30(8) = 5(48)$ $\qquad\qquad\qquad\qquad$ $-30(-8) = 5(48)$
$240 = 240$ ✓ $\qquad\qquad\qquad\qquad$ $240 = 240$ ✓

1. $\dfrac{84}{96} = \dfrac{7}{x}$

2. $\dfrac{120}{135} = \dfrac{x}{18}$

3. $\dfrac{301}{516} = \dfrac{7}{x}$

4. $\dfrac{x+8}{156} = \dfrac{x}{60}$

5. $\dfrac{32}{x} = \dfrac{24}{x-6}$

6. $\dfrac{x}{x+1} = \dfrac{x+6}{x+10}$

7. $\dfrac{x-5}{x} = \dfrac{x+5}{30}$

8. $\dfrac{4}{x-6} = \dfrac{x}{18}$

9. $\dfrac{x}{40} = \dfrac{3}{x-7}$

10. $\dfrac{5}{x-12} = \dfrac{x}{32}$

11. A 2-foot sunflower casts a 3-foot shadow on the ground. Nearby, a tree casts a 45-foot shadow. How tall is the tree?

12. Two numbers are in the ratio of 3:4. The smaller is 19 less than the larger. Find the numbers.

Percent

The term *percent* comes from Latin and means "out of one hundred." A percent measurement compares some part of a whole to some number out of one hundred. A percent is, therefore, a proportion.

$$\dfrac{\text{Part}}{\text{Whole}} = \dfrac{\text{Percent}}{100}$$

Consider the following sentence, for it is the basis of understanding all word problems involving percent.

The part is some percent of the whole.

Now let us diagram that sentence:

The part is some percent of the whole.

If we replace the words *the part, some* and *the whole* with numbers and insert the numbers into the proportion above, we have an equation.

157

Five is twenty-five percent of twenty.

$$\frac{5}{20} = \frac{25}{100}$$

The truth of the above equation can be determined in two ways. First, each fraction can be reduced to lowest terms.

$$\frac{1}{4} = \frac{1}{4}$$

Or instead, by cross-multiplication, the products of the means and extremes can be shown to be equal.

$$(20)(25) = (5)(100)$$
$$500 = 500$$

This brings us to using algebra to solve basic percent problems.

$$\frac{\text{Type I (“what is”)}}{\text{Type III (“of what”)}} = \frac{\text{Type II (“what %”)}}{100}$$

In Type I problems, the part, or "is", is unknown and must be found; the percent and the whole are given. A variable, x, is used in the proportion to replace the phrase, "what is." The variable always is placed in the numerator of the first fraction, above the whole.

Ex. 1a.   What is 25% of 20?

Diagramed, this becomes

What is 25% of 20?

$$\frac{x}{20} = \frac{25}{100}$$

Cross multiplying, we obtain

$(20)(25) = 100x$
$500 = 100x$
$x = 5$

There is a commonly used shortcut to solving Type I problems. If we take the proportion

$$\frac{x}{20} = \frac{25}{100}$$

and transform the entire percent fraction on the right side of the equals sign into a decimal, we obtain

$$\frac{x}{20} = 0.25 \quad \text{(Since 25 divided by 100 equals 0.25)}$$

Now, by cross multiplication, we multiply the whole by the decimal equivalent of 25% in order to obtain the part.

$$20(0.25) = x = 5$$

In Type II problems, the percent is unknown and must be found. The variable is used in the proportion to replace the phrase, "what percent." The variable always is placed in the numerator of the second fraction, above one hundred.

Ex. 1b. Five is what percent of 20?

Diagramed, this becomes

Five is what percent of 20?

$$\frac{5}{20} = \frac{x}{100}$$

Cross multiplying, we obtain
$$20x = 5(100)$$
$$20x = 500$$
$$x = 25$$

In Type III problems, the whole is unknown and must be found. The variable is used in the proportion to replace the phrase, "of what." The variable always is placed in the denominator of the first fraction, under the part. Type III problems are easily recognized by the awkwardness of their phrasing.

Ex. 1c. Five is 25% of what?

159

Diagramed, this becomes

Five is 25% of what?

An even more awkward phrasing is

Of what is five 25%?

Diagramed, this becomes

Of what is five 25%?

$$\frac{5}{x} = \frac{25}{100}$$

Cross multiplying, we obtain

25x = 5(100)
25x = 500
x = 20

1. What is 62.5% of 112?

2. What is 65% of 20?

3. What is 25% of 120?

4. What is 130% of 420?

5. 48 is what percent of 72?

6. 210 is what percent of 280?

7. 124 is what percent of 310?

8. 55 is what percent of 50?

9. 116 is 32% of what?

10. 84 is 42% of what?

11. Of what is fifteen 40%?

12. Of what is two hundred seventy-six 240%?

Tax is collected on sale items. Tax is a Type I problem; the additional step is to add sales tax to the cost of the item.

Ex. 2.   A dress costs $69. Sales tax is 8.625%. What is the total cost of the dress?

Solution: We find the dollar amount of the sales tax in either of the following two ways, and then add the amount to $69. A third option performs both steps in one.

Option 1	Option 2	Option 3

Option 1

$$\frac{x}{69} = \frac{8.625}{100}$$

$$x = \frac{(69)(8.625)}{100}$$

$$x = 5.95125 \approx 5.95$$

$$\begin{array}{r} 69 \\ + \quad 5.95 \\ \hline \$\ 74.95 \end{array}$$

Option 2

$$\begin{array}{r} .08625 \\ x \quad 69 \\ \hline 77625 \\ 51750 \\ \hline 5.95125 \end{array}$$

$$\begin{array}{r} 69 \\ + \quad 5.95 \\ \hline \$\ 74.95 \end{array}$$

Option 3

69(1.08625)
= 74.95125
≈ $ 74.95

The third option is based on the distributive property:

69 (1 + .08625)

= 69 + 5.95125

The decimal 1.0 equals 100%.

Discount is a percent reduction in the sales price. As a result, *discount is always subtracted* from the sale price. When both discount and tax are elements of a sales item, the discount is first found and subtracted from the whole. Then tax is calculated on the discounted whole and added to it.

Ex. 3.   The $ 69 dress is on sale at a 30% discount. The tax is 4.25 percent. What is the total cost of the dress?

Again, there are three options for obtaining the solution.

Option 1	Option 2	Option 3
		69(.7)
$\dfrac{x}{69} = \dfrac{30}{100}$	69	= 48.3
	x .3	
	20.7	48.3(1.0425)
$x = \dfrac{(69)(30)}{100}$		
	69.00	= 50.35275
	− 20.70	
x = 20.7 = 20.70	$ 48.30	≈ $ 50.35

69.00	48.3(1.0425)	The first step of the third option is based on the distributive property:
− 20.70		
$ 48.30	= 50.35275	
48.3(1.0425)	≈ $ 50.35	69 (1 − .3)
= 50.35275		= 69 − 20.7
≈ $ 50.35		The decimal 1.0 equals 100%.

Commission is a Type I problem whereby the salesperson receives a percentage of the sales price in exchange for her service. Because the commission comes out of the sale price of the item, the commission must be subtracted from the sale price.

Ex. 4.   A house sells for $450,000. The realtor earns 6% commission on the sale. How much is the realtor's commission and how much does the owner receive on the sale after paying the commission?

Solution: 450,000(.06) = $27,000,   450,000 − 27,000 = $423,000
Answer: The realtor's commission is $27,000; the owner receives $423,000 on the sale.

For problems 13-15, calculate the final cost of the item.
13.   Shoes: $89   Discount: 30%   Tax: 4.25%
14.   Coat: $400   Discount: 50%   Tax: 8.5%
15.   Suit: $159   Discount: 40%   Tax: 7%

For problems 16 and 17, calculate the commission on the price of the item.
16.   Car: $25,000   Commission: 3%
17.   House: $395,000   Commission 6%

Solving Single Variable Radical Equations

The strategy for solving all elementary radical equation is to isolate the radical expression by applying additive and/or multiplicative inverses in necessary order. The student must identify whether the radical is attached to an addition or multiplication chain in order to determine which inverse to apply. Once the radical is isolated, the entirety of both sides of the equation are squared; this frees the radicand and permits for the variable to continue to be isolated by means of inverses in as many steps as necessary until the value of the variable is found.

Ex. 1. $\sqrt{x} = 6$ Square both sides; this removes the radical because
$(\sqrt{x})^2 = 6^2$ the square root of the radicand squared equals the radicand.
$x = 36$

Ex. 2. $\sqrt{\dfrac{x}{4}} = 5$ Square both sides; this removes the radical.

$\dfrac{x}{4} = \dfrac{25}{1}$ Multiply the means (every whole number has an invisible
          denominator of 1).

$x = 100$

Ex. 3.

$15\sqrt{\dfrac{4}{x}} = 75$ Unlink 15 with its multiplicative inverse.

$\sqrt{\dfrac{4}{x}} = 5$ Square both sides.

$\dfrac{4}{x} = 25$ Switch the means (every whole number has an invisible denominator of
          one).

$\dfrac{4}{25} = x$

1. $\sqrt{x} = 11$

2. $\sqrt{x-3} = 7$

3. $-8 = \sqrt{x-3}$

4. $1 + \sqrt{2a-3} = 6$

163

5. $\sqrt{x-5} - 20 = 13$

6. $1 + \sqrt{15x-14} = 3$

7. $3\sqrt{x-4} = 27$

8. $3\sqrt{4x} = 30$

9. $7\sqrt{\dfrac{x}{2}} = 84$

10. $11\sqrt{\dfrac{36}{x}} = 22$

11. $\dfrac{\sqrt{x-3}}{2} = 2$

12. $\sqrt{x+12} = \sqrt{8}$

Solving Fractional Equations

The best strategy for solving fractional equations is to multiply every fraction on the left and right side of the equation by the lowest common denominator. This eliminates the fractions from the equation, producing a simple linear or quadratic equation for solving.

Ex. 1.   Solve for x:

$$5 = \frac{2x - 12}{2x + 4} + \frac{x - 10}{2x + 4}$$

Solution: The fractions on the right have the same denominator; 5 on the left has an invisible denominator of one. We multiply every term on the left and right side of the equals sign by $2x + 4$:

$$5(2x + 4) = \frac{(2x - 12)(2x + 4)}{2x + 4} + \frac{(x - 10)(2x + 4)}{2x + 4}$$

On the right, the factors of $2x + 4$ cancel one-to-one in numerator and denominator; 5 is distributed to $2x + 4$ on the left.

$10x + 20 = 2x - 12 + x - 10$ Combine like terms.

$10x + 20 = 3x - 22$ Unlink 3x from the right; unlink 20 from the left.

164

$7x = -42$ Unlink 7 with its multiplicative inverse.
$x = -6$

Check:

$$5 = \frac{2(-6) - 12}{2(-6) + 4} + \frac{(-6) - 10}{2(-6) + 4}$$

$$5 = \frac{-12 - 12}{-12 + 4} + \frac{-16}{-12 + 4}$$

$$5 = \frac{-24}{-8} + \frac{-16}{-8}$$

$$5 = 3 + 2$$
$$5 = 5 \checkmark$$

A common error that students make is to cross-multiply. Cross-multiplication is not possible because there are *two* or more fractions on one or both sides of the equals sign connected by addition or subtraction signs; because of this, a proportion does not exist. While it is be possible to give the fractions on each side their common denominators, combine those fractions into one on each side and convert the equation to a proportion, this compounds variables, creating quadratic equations with extraneous solutions or higher degree equations which cannot be solved at the basic algebra level.

1. $\dfrac{1}{x} + \dfrac{1}{x^2} = \dfrac{4}{x}$

2. $\dfrac{1}{x^2} - \dfrac{1}{x} = \dfrac{4}{x}$

3. $\dfrac{11}{x^2} = \dfrac{3}{x} - \dfrac{1}{x^2}$

4. $\dfrac{x + 12}{3x} = \dfrac{-x - 2}{x} - \dfrac{x + 2}{3x}$

5. $1 = \dfrac{12 - 4x}{2x - 8} + \dfrac{4}{2x - 8}$

165

APPLICATIONS IN TWO VARIABLES:
X AND *Y* AS CHANGING VALUES

In this chapter we will explore equations in which x no longer necessarily represents an unknown; instead, x and y are linked together in an equation representing their relationship with respect to each other. As x changes in value, y changes in value, too. The changing values of x and y as defined by a unique equation create graphs of lines and curves in the coordinate plane. Before we explore these equations and graphs, we will first explore some common equations containing variables whose values are all known except for one, and learn how to isolate the one unknown variable when solving for it.

__Expressing One Variable in Terms of the Other(s)__

Consider the common Celsius -to-Fahrenheit temperature conversion formula

$$F = \frac{9}{5} C + 32$$

The formula works perfectly as written if we know the Celsius temperature and wish to know its Fahrenheit equivalent. So, for example, if we are in Paris and are told that it is $15\,°C$, we substitute 15 for the variable C to find that

$$F = \frac{9}{5} (15) + 32 = 9(3) + 32 = 27 + 32 = 59\,°F$$

We say that F *is expressed in terms of C*, meaning that F derives its value and meaning based on the numerical manipulations being performed on C.

C is first multiplied by 9 and divided by 5, or divided first by 5 and multiplied by 9 (possible to do because multiplication is associative); that result is then added to 32.

Suppose, however, that a Parisian in New York is told that it is $86\,°F$ and wishes to know the Celsius equivalent. The formula as written is inconvenient to use; we want *C expressed in terms of F*.

Examining the formula, we analyze the operations performed on C in order to unlink them and apply inverses in the correct order. Since the last step for evaluating C is to add 32, we unlink addition of 32 with subtraction of 32 to both sides of the equation.

$$F = \frac{9}{5}C + 32$$

$$\frac{-32 \qquad -32}{F - 32 = \frac{9}{5}C}$$

A short subtraction chain now exists on the left side of the equation. This subtraction chain must now be treated as a unity of highest priority in the order of operations and protected by parentheses.

$$(F - 32) = \frac{9}{5}C$$

The last step is to unlink multiplication of C by 9/5 with multiplication of 5/9 to both sides of the equation.

$$\frac{5}{9}(F - 32) = C$$

We now have an equation which expresses Celsius temperature in terms of Fahrenheit temperature. Substituting the New York temperature of 86F into the equation and following the order of operations, we obtain

$$\frac{5}{9}(86 - 32) = \frac{5}{9}(54) = 5(6) = 30\,C$$

Regardless of the number of variables to be found in any given formula, the procedure for isolating one of those variables is to examine the operations connecting them and to perform the inverses of those operations to both sides. This results in expressing any variable in terms of the others.

1. Express the length of a rectangle in terms of area and width.
    $$A = lw$$

2. Express the radius of a circle in terms of circumference.
    $$C = 2\pi r$$

3. Express b_1 of a trapezoid in terms of the other variables.
    $$A = \frac{(b_1 + b_2)h}{2}$$

The Coordinate Plane

The axes of the coordinate plane are simply two number lines running from negative infinity to positive infinity and intersecting at a 90°angle. The horizontal axis is called the *x-axis* while the vertical axis is called the *y-axis*. The axes intersect at the point which is zero on both axes; this point, whose (x, y) coordinates are (0, 0), is called the *origin*. The *coordinate plane* or *Cartesian plane* is the two-dimensional flat surface defined by this system. By means of this system, any location or position with respect to the origin can be identified by a pair of horizontal and vertical coordinates, both numerical, written as (x, y) respectively.

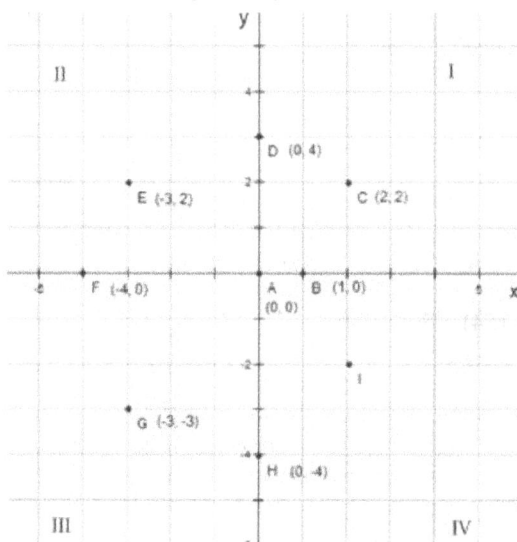

The axes create four regions of the plane known as *quadrants*. Roman numerals I, II, III and IV number each quadrant beginning with the upper right region and moving counterclockwise as illustrated above. Point A is the origin (0, 0). Point B is on the positive x-axis and is labeled (1, 0) because the point is one unit right of the origin but moves neither up nor down. Point C is in the first quadrant and is labeled (2, 2) because its position is two units right of the origin and two units above the horizontal position. Point D, located on the positive y-axis, has coordinates (0, 4) because it moves neither right nor left of the origin but only up four units. Point E, located at (– 3, 2) in Quadrant II, is three units left of the origin and two units above its horizontal position. Point F located on the negative x-axis, has coordinates (– 4, 0) because it moves four units left of the origin but neither up nor down.

168

Point G with coordinates (– 3, – 3) moves three units left of the origin and three units down from its horizontal position.

Point H with coordinates (0, – 4) moves neither right nor left of the origin but only four units down the negative y-axis. Contrast the coordinates of Point H with those of Point F to realize that interchanging x- and y- coordinates indicates an entirely different position on the plane.

Notice that every point on the x-axis has zero as its y-coordinate because there is no vertical movement off the axis after the initial horizontal movement left or right of zero, and that every point on the y-axis has zero as its x-coordinate because there is no initial horizontal movement left or right of the origin before moving up or down the y-axis.

The General Form of a Linear Equation

The general form of a linear equation is

$$Ax + By = C \quad \text{or} \quad Ax + By + C = 0$$

where A, B and C are numerical *constants* of the equation and x and y are the *variables*, or changing values, of the equation. The numerical values of A, B and C determine the direction and position of the line on the coordinate plane; as a result, A, B and C also determine the relationship of x and y with respect to each other. The coordinates of all points (x, y) on the line change as one moves from point to point on the line; in this context, the definition of variable is not so much an unknown but rather a changing value or set of x- and y-values depending upon which point on the line we are on. The set of all points (x, y) will make the equation true when both coordinates are substituted into the equation and thus will *satisfy the equation*. Any point not on the line will make the equation false when those coordinates are substituted into the equation and thus will fail to satisfy the equation.

Before we look at a specific example, notice the format of the equation. The left side contains an addition link between two short multiplication chains. This observation will be important when solving for either x or y.

Consider the equation

$$x + y = 4$$

The invisible coefficients are A = 1 and B = 1; C = 4. To find the points where the line crosses the x- and y-axes, we simply set each variable equal to zero, one at a time, to discover the value of the other coordinate.

When x = 0, then $\qquad$ $0 + y = 4$
$$y = 4$$

 The ordered pair is (0, 4). The y-coordinate of the line's point on the y-axis is called the *y-intercept*. Plot the point.

When y = 0, then $\qquad$ $x + 0 = 4$
$$x = 4$$

The ordered pair is (4, 0). The x-coordinate of the line's point on the x-axis is called the *x-intercept*. Plot the point.

To find other points on the line, choose random values of x on both sides of zero and substitute them into the equation to solve for y. Plot the points.

x	x + y = 4	Unlink addition link with inverse to both sides	y
− 2	− 2 + y = 4	y = 4 + 2	6
1	1 + y = 4	y = 4 − 1	3
6	6 + y = 4	y = 4 − 6	− 2

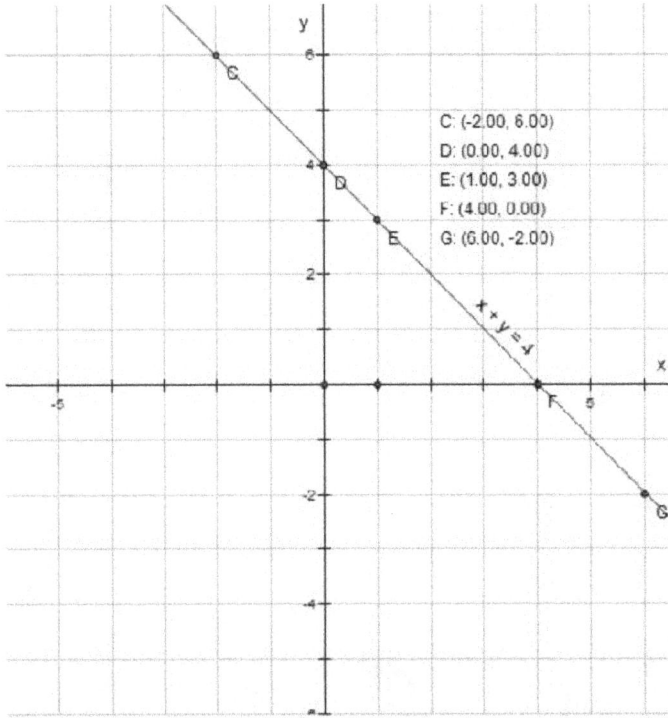

C: (-2.00, 6.00)
D: (0.00, 4.00)
E: (1.00, 3.00)
F: (4.00, 0.00)
G: (6.00, -2.00)

The points are connected with a straight edge to form the line. Note that only two points are needed in order to determine the line; thus, the x- and y-intercepts are sufficient to graph the line.

The procedure is only slightly more complicated when A and B have values other than (invisible) one.

To graph the line

$$x - 2y = 5$$

we use the same procedure for finding the x- and y-intercepts, but this time we include them in the table.

171

x	x – 2y = 5	(1) Unlink Addition Link with inverse to both sides (2) Unlink Multiplication Link with inverse to both sides	y
5	substitute y = 0 here $x - 2(0) = 5$	$x - 0 = 5, \quad x = 5$	0
0	$0 - 2y = 5$	$-2y = 5, \quad y = 5/-2$	$-5/2$
-1	$-1 - 2y = 5$	$-2y = 5 + 1, \quad y = 6/-2$	-3
1	$1 - 2y = 5$	$-2y = 5 - 1, \quad y = 4/-2$	-2

We plot the points.

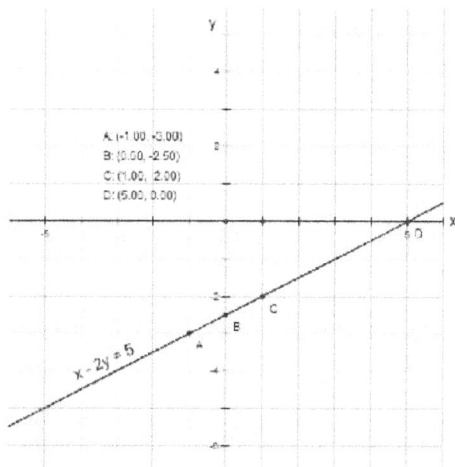

A: (-1.00, -3.00)
B: (0.00, -2.50)
C: (1.00, -2.00)
D: (5.00, 0.00)

x – 2y = 5

In the next example, all terms are on the left side of the equation and zero is on the right side, we unlink C from the addition chain and move its additive inverse to the right.

$$3x - 4y + 6 = 0 \quad \text{becomes} \quad 3x - 4y = -6$$

x	3x – 4y = –6	(1) Unlink Addition Link with inverse to both sides (2) Unlink Multiplication Link with inverse to both sides	y
–2	$3x - 4(0) = -6$	$3x = -6, \; x = -2$	0
–1	$3(-1) - 4y = -6$	$-3 - 4y = -6, \; -4y = -3$	$^3/_4$
0	$3(0) - 4y = -6$	$-4y = -6, y = -6/-4$	$^3/_2$
1	$3(1) - 4y = -6$	$3 - 4y = -6, -4y = -9$	$^9/_4$

We plot the points.

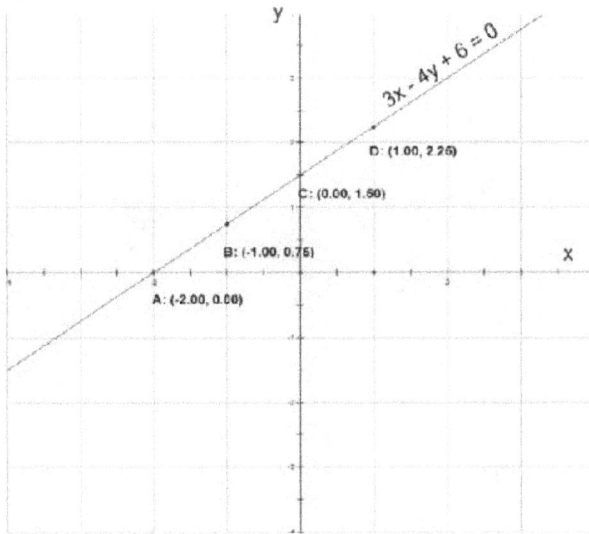

Try graphing the following equations on graph paper.

1. $2x + 3y = 6$
2. $3x - 2y = 8$
3. $2x + 4y = 5$
4. $5x - 4y + 4 = 0$
5. $3y - 6x - 9 = 0$

The Slope Formula

The slope of a line determines the direction of a line in the plane. Whether a line has a gentle slope or a very steep slope and whether it rises or falls as one moves from left to right in the plane boils down to a simple numerical comparison of the number of units that the vertical coordinate changes up or down to the number of units that the horizontal coordinate advances right as you move from point to point on the line. Consider the equation

$$y - x = 0$$

If we isolate y in order to express it in terms of x, we move the additive inverse of $-$ x to the opposite side to obtain

$$y = x$$

While a table could be used, this simple equation states that the value of y is whatever the value of x is. Thus, $(-3, -3), (-1, -1), (0, 0), (2, 2)$ and $(4, 4)$ are plotted on the plane.

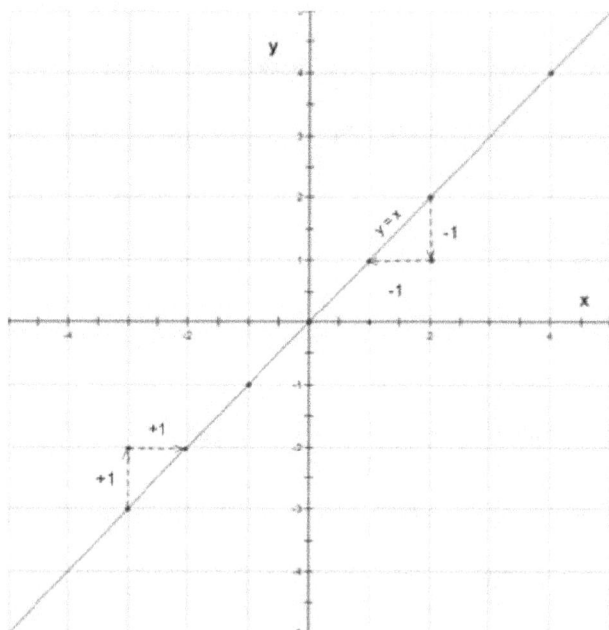

174

Moving from $(-3, -3)$ to $(-2, -2)$ on the line, we note that y increases by one unit as x increases by one unit. Since a comparison of two quantities is expressed as a ratio in mathematics, we write,

$$m = \frac{\Delta y}{\Delta x} = \frac{1}{1} = 1$$

where m represents the slope of the line, and the Greek letter *Delta* (Δ) represents the change. This, however, does not complete the formula; mathematically, a change in a quantity is a difference between two values, and the term *difference* implies *subtraction*. Thus we expand the formula:

$$m = \frac{\Delta y}{\Delta x} = \frac{y_2 - y_1}{x_2 - x_1}$$

where y_2 represents the y-coordinate of the second point, y_1 represents the y-coordinate of the first point, x_2 represents the x-coordinate of the second point and x_1 represents the x-coordinate of the first point. Substituting $(-3, -3)$ for (x_1, y_1) and $(-2, -2)$ for (x_2, y_2) yields the same result.

$$m = \frac{\Delta y}{\Delta x} = \frac{y_2 - y_1}{x_2 - x_1} = \frac{-2 - -3}{-2 - -2} = \frac{1}{1} = 1$$

The steeper the line, the steeper the slope; this geometric observation is confirmed algebraically. Given the equation $y = 3x - 5$, we substitute $\{-1, 0, 1, 2\}$ for values of x to find the corresponding y-coordinates and plot the points.

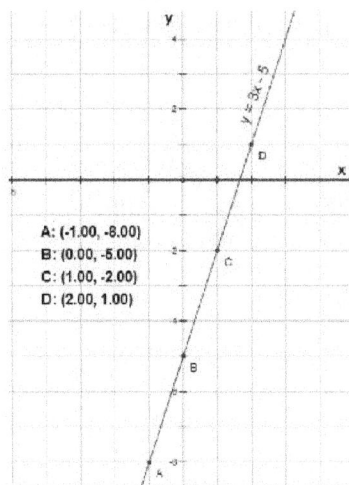

A: (-1.00, -8.00)
B: (0.00, -5.00)
C: (1.00, -2.00)
D: (2.00, 1.00)

$$m = \frac{\Delta y}{\Delta x} = \frac{y_2 - y_1}{x_2 - x_1} = \frac{-5 - -8}{0 - -1} = \frac{-5 + 8}{0 + 1} = \frac{3}{1} = 3$$

Notice that the slope of 3 corresponds to the coefficient of the x-term in the equation. The coefficient of x is always, in fact, the slope of the line whenever y is isolated on one side of the equation with an invisible coefficient of + 1.

The slope of a line may be fractional and/or negative, as in the next example.

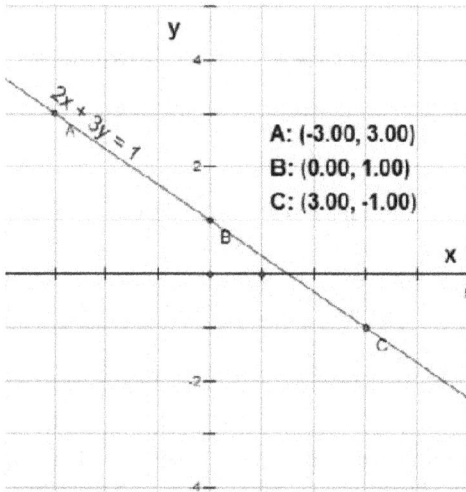

A: (-3.00, 3.00)
B: (0.00, 1.00)
C: (3.00, -1.00)

To move from the point (– 3, 3) to (0, 1), the vertical distance is 2 units down and the horizontal distance is 3 units right. This geometric observation is confirmed algebraically with the slope formula.

$$m = \frac{\Delta y}{\Delta x} = \frac{y_2 - y_1}{x_2 - x_1} = \frac{1 - 3}{0 - -3} = \frac{-2}{0 + 3} = \frac{-2}{3}$$

A common error is to write the difference of the x-coordinates in the numerator and the difference of the y-coordinates in the denominator. This simply confuses the order of the ordered pair (x first, y second) with the order of differences in the slope formula (difference of y-coordinates above, difference of x-coordinates below).

Find the slope between the given points.

1. (1, 1) and (4, 5)   2. ($^-$3, $^-$2) and (3, 2)   3. (2, 5) and ($^-$4, $^-$5)

The Slope-Intercept Form of a Linear Equation

Consider the equation

$$-3x + y = 2$$

If we move the additive inverse of $-3x$ to the other side of the equation, we obtain

$$y = 3x + 2$$

We select values of -1, 0 and 1 for x to find y:

$$y = 3(-1) + 2 = -3 + 2 = -1$$
$$y = 3(0) + 2 = 0 + 2 = 2$$
$$y = 3(1) + 2 = 3 + 2 = 5$$

The ordered pairs are $(-1, -1)$, $(0, 2)$ and $(1, 5)$. We plot the points.

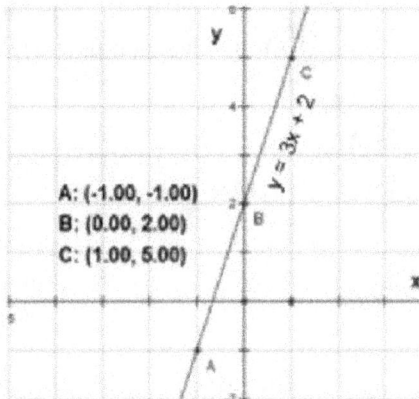

The slope of the line can be determined by substituting the coordinates of any two points on the line into the slope formula. We use Points A and B.

$$m = \frac{\Delta y}{\Delta x} = \frac{y_2 - y_1}{x_2 - x_1} = \frac{2 - -1}{0 - -1} = \frac{2 + 1}{0 + 1} = \frac{3}{1} = 3$$

This matches the graphical result of moving up three units and right one unit from point to point. But the coefficient of the x term in the equation is also three. In fact,

$$y = 3x + 2$$
$$y = mx + b$$

is a linear equation in *slope-intercept* form. The other numeral is equally significant: two is the y-intercept of the equation, that is, the y-coordinate of the point where the line crosses the y-axis. Whereas the slope of a line determines its direction, the y-intercept of a line determines its position on the plane.

The slope-intercept form of a linear equation provides for the fastest way to graph a linear equation. By rewriting any linear equation from general to slope-intercept form, the y-intercept is identified and plotted first, and then the slope, the multiplier of x, is interpreted as a quotient, a ratio, in order to plot points left and right of the y-intercept. Given

$4x + 6y = -6$	$Ax + By = C$	Unlink Ax with its I additive inverse
$6y = -4x - 6$	$By = -Ax + C$	Unlink B with its multiplicative inverse.
$y = \dfrac{-4}{6}x - 1$	$y = \dfrac{-A}{B}x + \dfrac{C}{B}$	Reduce to simplest form.
$y = \dfrac{-2}{3}x - 1$		

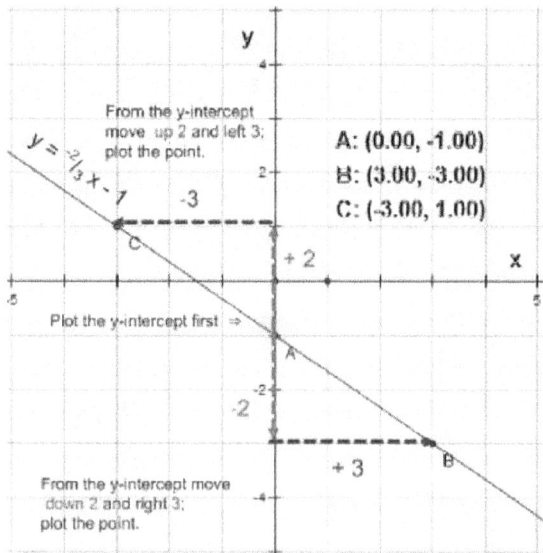

From the y-intercept move up 2 and left 3; plot the point.

$y = -\frac{2}{3}x - 1$

-3

C.

A: (0.00, -1.00)
B: (3.00, -3.00)
C: (-3.00, 1.00)

+ 2

Plot the y-intercept first ⇒

A

-2

+ 3

B

From the y-intercept move down 2 and right 3; plot the point.

178

By comparing an equation in both forms, we derive a formula showing the relationship between the constants in both forms in order to find the slope and y-intercept without having to rewrite the equation.

$$y = mx + b$$
$$y = \frac{-A}{B}x + \frac{C}{B}$$

Thus, $m = \frac{-A}{B}$ and $b = \frac{C}{B}$

Note that the y-intercept equals the constants C/B from the general form of the equation when C is already isolated on the right side; when it is not, then the y-intercept is – C/B because C, like A, would have its additive inverse moved to the right side.

In the first example in this subsection, the slope is three. Integral slopes are always interpreted as a quotient with an invisible denominator of one; the integer is the vertical movement up or down, and the invisible one is the horizontal movement right.

Identify the slope and y-intercept without rewriting the equation.

1. $3x - 2y = 6$
2. $x - 2y = 4$
3. $6y - 4x - 8 = 0$

Rewrite the following equations in slope-intercept form and graph.

4. $2x - 3y = 9$
5. $3x + 4y = -8$
6. $4x - 3y + 12 = 0$

Graphing Lines of the Form y = b

Consider the points on the line $y = -5$ graphed below.

The y-coordinate of every point on the line is -5 without variation; hence, the name $y = -5$.

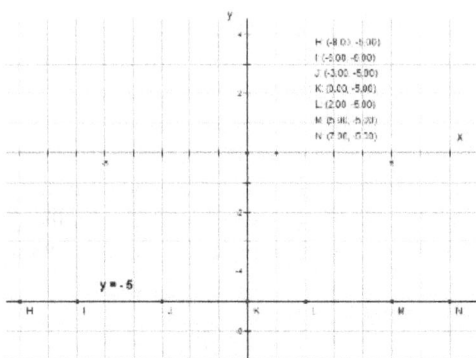

When we apply the slope formula to any two points on this or any other horizontal line, the value of the numerator is always zero. We use $(1, -5)$ and $(-3, -5)$ as an example to substitute into the formula.

$$m = \frac{\text{rise}}{\text{run}} = \frac{y_2 - y_1}{x_2 - x_1} = \frac{-5 - ^-5}{-3 - 1} = \frac{0}{-4} = 0$$

The slope of every horizontal line is zero because the line neither inclines up nor declines down but remains steady. Every horizontal line is called a *constant function* because it is of the form y = b where b is an element of the set of real numbers. This equation is derived from the slope-intercept form of a linear equation when zero is substituted for *m* into the equation.

$$y = mx + b$$
$$y = 0x + b$$
$$y = b$$

Graph the following equations.

1. $y = -3$ 2. $y = 5$ 3. $y = 0$

180

Graphing Lines of the Form x = a

Consider the points on the vertical line x = 4 below.

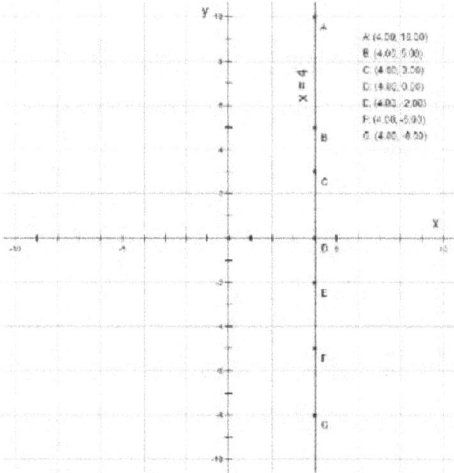

The x-coordinate of every point on the line is 4 without variation; hence, the name x = 4.

When we apply the slope formula to any two points on this or any other vertical line, the value of the denominator is always zero. We use (4, 6) and (4, –2) as an example to substitute into the formula.

$$m = \frac{rise}{run} = \frac{y_2 - y_1}{x_2 - x_1} = \frac{6 - {}^-2}{4 - 4} = \frac{8}{0} = \text{undefined}$$

Since division by zero is mathematically impossible, we say that the slope of any vertical line with formula x = a is *undefined*. The term *undefined* expresses the idea that no numerical value for the slope can be obtained because the slope increases without bound in the positive y direction and decreases without bound in the negative y direction. Vertical lines can also be said to have *no slope, no defined slope* or *infinite slope*.

Graph the following equations.

1.   x = 7   2.   x = – 4   3.   x = 0

The Point-Slope Form of a Linear Equation

The point-slope form of a linear equation is derived from the slope formula.

$$m = \frac{y_2 - y_1}{x_2 - x_1}$$

Recall that there is an implied set of parentheses around the numerator and denominator of the slope quotient. If we multiply both sides of the equation by $x_2 - x_1$, we obtain

$$m(x_2 - x_1) = y_2 - y_1$$

Bear in mind that the subscript numerals indicate a constant, and not a changing, value for the x- and y-coordinates to which they are attached. We replace the constants x_2 and y_2 with the variables x and y to obtain the formula.

$$y - y_1 = m(x - x_1)$$

Both the point-slope and slope-intercept forms give the slope of the line which can easily be read from the equation. The equations differ in that the slope-intercept form gives the y-intercept, that is, the y-coordinate of the point where the line crosses the y-axis, whereas the point-slope equation gives any point on the line. While the slope-intercept form is most convenient for graphing equations, the point-slope form is more flexible for writing equations because any point on the line can be substituted for (x_1, y_1) in order to write the equation.

Writing the Equation of a Line Given the Slope and One Point

The point-slope form is most convenient for writing the equation of a line given the slope of the line and any point on it. Simply substitute into the equation and you are done.

Ex. 1.   Write the equation of a line with a slope of 3 and passing through the point (2, 4).

Solution: We substitute m = 3 and (− 2, 4) for (x_1, y_1) into

$$y - y_1 = m(x - x_1)$$

$$y - 4 = 3(x - - 2)$$
$$y - 4 = 3(x + 2)$$

Ex. 2.   Write the equation of a line with a slope of − 4 passing through the point (− 2, − 3) in slope-intercept form.

182

Solution: We substitute $m = -4$ and $(-2, -3)$ for (x_1, y_1) into

$y - y_1 = m(x - x_1)$
$y - -3 = -4(x - -2)$ Simplify the double negatives with **keep-plus-opposite**.
$y + 3 = -4(x + 2)$ Distribute -4 to the addition chain inside parentheses.
$y + 3 = -4x - 8$ Unlink 3 with its additive inverse.
$y = -4x - 8 - 3$ Combine like terms to identify the y-intercept.
$y = -4x - 11$

Write the equations of the lines in slope intercept form given the slope and one point.

1. $m = 3$, passing through $(8, -3)$
2. $m = -2$, passing through $(-1, -2)$
3. $m = \frac{2}{3}$, passing through $(-3, -4)$
4. $m = \frac{1}{4}$, passing through $(4, 3)$
5.   Write the equation of the line with a slope of zero and passing through $(2, 5)$.

Writing the Equation of a Line Given Two Points

When asked to do this, the slope is not given, but two points are, so find the slope between the two points first. The rest of the procedure is identical to that in the preceding section.

Ex. 1.   Write the equation of the line passing through $(-1, -1)$ and $(3, 4)$ in slope-intercept form.

Solution: We find the slope between the points first.

$$m = \frac{\Delta y}{\Delta x} = \frac{y_2 - y_1}{x_2 - x_1} = \frac{4 - -1}{3 - -1} = \frac{5}{4}$$

We use the coordinates of either point to substitute into the formula.

$y - y_1 = m(x - x_1)$
$y - -1 = \frac{5}{4}(x - -1)$ Simplify the double negatives with **keep-plus-opposite**.
$y + 1 = \frac{5}{4}(x + 1)$ Distribute $\frac{5}{4}$ to the addition chain inside parentheses.
$y + 1 = \frac{5}{4}x + 5/4$ Unlink 1 with its additive inverse.
$y = \frac{5}{4}x + 5/4 - 1$ Combine like terms to identify the y-intercept.
$y = \frac{5}{4}x + \frac{1}{4}$ Check the equation with the other point;
                                   this is left to the student.

183

In the following problems, write the equation of the line given two points.

1. (2, 3) and (0, 2)
2. (– 1, 3) and (4, – 2)
3. (– 2, – 5) and (1, – 3)
4. (– 2, 4) and (7, 4)
5. (2, – 3) and (4, – 1)

Solving Systems of Linear Equations in Two Variables

To find the solution of a system of linear equations means to find the point of intersection of two lines in the plane. Two lines will always intersect in a plane in exactly one point as long as they are (i) not parallel and (ii) non-concurrent, that is, one line superimposed on the other. The *solution* of the system is the ordered pair (x, y) where the lines intersect; the solution may be found graphically or by two separate algebraic methods. Only the algebraic method will be discussed in this subsection because it is often the only viable method for precisely solving systems of equations with messy, fractional y-intercepts. Regardless of the method used, the number of equations must always equal the number of variables, in this case two.

The first of the two algebraic methods is the *substitution* method, whereby a variable from one equation is expressed in terms of the other variable and substituted into the other equation in order to temporarily remove one variable.

Ex. 1. 1st: $\quad x + y = 1$ Rewrite the second equation to express x
$\quad\quad 2^{nd}$: $x - y = - 3$ in terms of y.

$\downarrow$

2nd: $x = y - 3$ Substitute y – 3 for x into the first equation and
1st: $x + y = 1$ reduce the system to one variable in one equation.

$(y - 3) + y = 1$ Note the addition chain; combine like terms.
$2y - 3 = 1$ Unlink – 3 with its additive inverse to both
$2y = 4$ sides; then unlink 2 with its multiplicative
$y = 2$ inverse to both sides.

      Substitute 2 for y into either equation in order to
      find x; unlink 2 with its additive inverse.

$x + (2) = 1$
$x = - 1$ Express the solution as an ordered pair; the check is
Ans: (– 1, 2)   left to the student.

184

The second method is known as the *addition* method. Like the substitution method, we temporarily remove one variable and reduce the system to one equation in order to solve for the other variable. With this method, the equations are added to one another such that the variable to be removed plus its additive inverse in the second equation equal zero.

Ex. 2. 1st: $x + y = 1$ Add the two equations: add the left side to the
2nd: $x - y = -3$ left side and the right side to the right side
                             Note that y minus y equals zero on the left
$2x = -2$ Unlink 2 with its multiplicative inverse to both
$x = -1$ sides.

$(-1) + y = 1$ Substitute -1 for x into either equation (here,
$y = 2$ the first); unlink -1 with its additive inverse.

Ans: $(-1, 2)$ Express the solution as an ordered pair. The check
                             is left to the student.

Sometimes the entirety of one equation must be multiplied by a constant in order to create the additive inverse of the variable to be removed.

Ex. 3. 1st: $\frac{1}{3}x + y = -1$ Multiply the first equation by -3
2nd: $-2x + 3y = -12$ Add the two equations: add the left side to the
                                 left side and the right side to the right side.
1st: $-x - 3y = 3$ Note that $-3y$ plus $3y$ equals zero on the left
2nd: $-2x + 3y = -12$
                                 Unlink -3 with its multiplicative inverse to both
                                 sides.
$-3x = -9$
$x = 3$ Substitute $x = 3$ into either equation to find y.

$\frac{1}{3}(3) + y = -1$
$1 + y = -1$
$y = -2$ Express the solution as an ordered pair. The check
Ans: $(3, -2)$ is left to the student.

Sometimes both equations must be multiplied by a constant before adding them will render one equation in one variable.

Ex. 4. 1st: $3x + 5y = 7$ Multiply the first equation by 5 and the second
2nd: $-5x + 3y = -23$ equation by 3.

185

$$15x + 25y = 35$$
$$-15x + 9y = -69$$

Add the two equations: add the left side to the left side and the right side to the right side. Note that 15x plus – 15x equals zero on the left.

$$34y = -34$$
$$y = -1$$

Unlink 34 with its multiplicative inverse. Substitute y = – 1 into either equation to find x.

$$3x + 5(-1) = 7$$
$$3x = 12$$
$$x = 4$$
Ans: $(4, -1)$

Express the solution as an ordered pair. The check is left to the student.

Solve the systems by the more convenient method.

1. $x + y = 1$
         $x - y = 3$

4. $x + y = 6$
         $-5x + y = -6$

2. $x + 3y = -3$
         $-2x + 3y = -12$

5. $-3x + 4y = 0$
         $x + 2y = -10$

3. $y = 2x + 5$
         $2x + y = -3$

6. $-2x + 3y = -6$
         $3x + 2y = -17$

Finding the solution of a horizontal and vertical line requires no algebraic manipulation; simply place the constant values for x and y into a set of parentheses. So the solution of x = 6 and y = 4 is (6, 4).

Graphing Systems of Linear Inequalities in Two Variables

The procedure for graphing a linear inequality is almost the same as that for graphing a linear equation, with a few extra steps. The line is rewritten in slope-intercept form, the y-intercept is plotted and the slope is interpreted as vertical-to-horizontal movements to get to the next point. The difference between an equation and an inequality is that only the points on the line satisfy the truth of the equation, whereas all points in the plane either above or below and perhaps on the line itself will satisfy the inequality. Given the inequality

$$y \leq \tfrac{3}{4} x + 2 \quad (\text{y is less than or equal to } \tfrac{3}{4} \text{ x plus 2})$$

every point on the line and all points below the line will satisfy the equation. The first graph below shows a solid line and the area below the line shaded in green as the complete solution set.

186

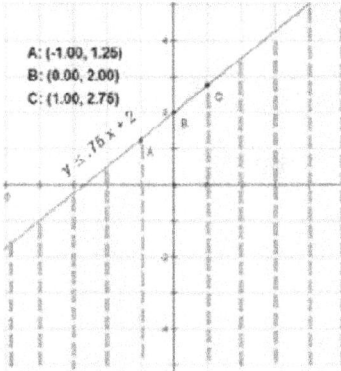

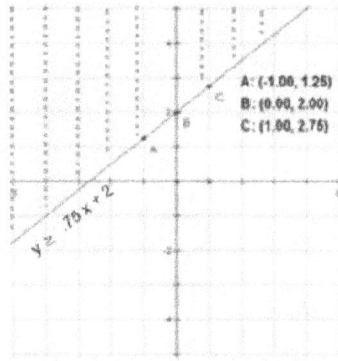

A: (-1.00, 1.25)
B: (0.00, 2.00)
C: (1.00, 2.75)

$y = .75x + 2$

A: (-1.00, 1.25)
B: (0.00, 2.00)
C: (1.00, 2.75)

$y = .75x + 2$

Contrast

$$y \geq \tfrac{3}{4} x + 2 \qquad \text{(y is greater than or equal to } \tfrac{3}{4} \text{ x plus 2)}$$

which shows a solid line and the area above the line shaded in green as the complete solution set.

The solid line, used when the symbol is $\leq$ or $\geq$, indicates that every point on the line is *part* of the solution set, that is, the coordinates of every point on the line will satisfy the inequality when substituted into it. A dashed line is drawn when the inequality symbol is either less than (<) or greater than (>). When either of these symbols are used, the line is only a boundary for and not part of the solution set. The area above the line is shaded when the symbol is greater than (>) and the area below the line is shaded when the symbol is less than (<). If the student is unsure which part of the plane is above or below a line, the coordinates of any test point clearly to one side of the line can be substituted into the inequality. If the test point satisfies the inequality, the area containing the test point is shaded; if the test point fails to satisfy the inequality, the other side of the line is shaded. This procedure will be illustrated in the next example.

Graphing a system of inequalities involves graphing the two lines, either solid or dashed, with their shaded areas on the same plane.

To graph the solution set of the system

$$y < 2x + 2$$
$$y \geq -3x + 3$$

187

each inequality is graphed one at a time. For each inequality:

i.   Plot the y-intercept;

ii. Substitute – 1, and 1 for x to find the corresponding y-values and plot the ordered pairs, or use the slope to plot a point left and right of the y-intercept;

iii. Connect the points with a solid line if the symbol is ≤ or ≥; use a dashed line if the symbol is < or >;

iv. Shade above the line for > or ≥ or below the line for < or ≤.   In the alternative, substitute the coordinates of a test point into the inequality.

Test (0, 0):	$y < 2x + 2$	$y \geq -3x + 3$
	$0 < 2(0) + 2$	$0 \geq -3(0) + 3$
	$0 < 2$ ✓	$0 \geq 3$ ✘
	Shade the side	Fails; shade the side
	containing (0, 0).	that does not contain
		(0, 0).

v.   Label the cross-hatched area with the words "solution set."

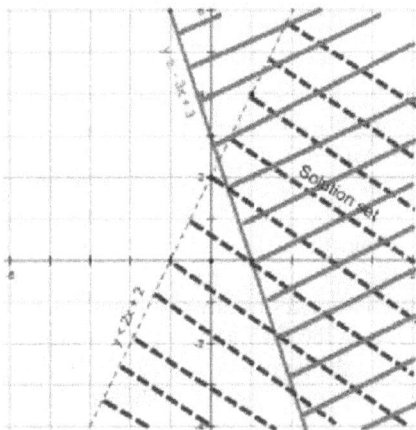

Bear in mind that systems of inequalities can only be solved graphically. There is no algebraic solution because the inequality symbols thwart the theory and mechanisms which make addition of equations possible.

Graph the following systems of equations as exercises.

1. $y \geq 2x - 2$ 2. $y \leq -2x + 3$

       $y < \frac{1}{2}x + 1$ $y > -3$

Quadratic Equations and the Parabola

Consider the equation
$$y = x^2$$

If we select values of x from -3 to $+3$ and substitute them into the equation, we obtain

x	y
-3	9
-2	4
-1	1
0	0
1	1
2	4
3	9

Plotting the ordered pairs on the coordinate plane, we obtain the graph of a parabola.

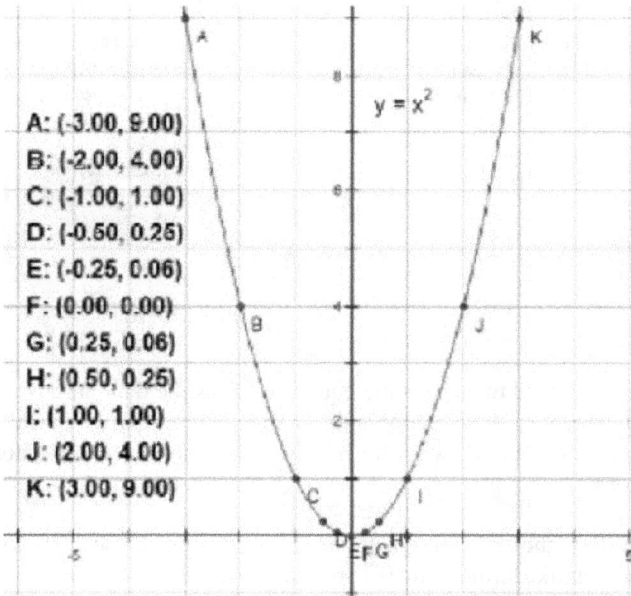

A: (-3.00, 9.00)
B: (-2.00, 4.00)
C: (-1.00, 1.00)
D: (-0.50, 0.25)
E: (-0.25, 0.06)
F: (0.00, 0.00)
G: (0.25, 0.06)
H: (0.50, 0.25)
I: (1.00, 1.00)
J: (2.00, 4.00)
K: (3.00, 9.00)

$y = x^2$

The graph is *u*-shaped; a common error is to connect the points in the shape of a *v*. Notice the ordered pairs with fractional values of x between -1 and

+1, labeled D through H; we see that as x-values get closer to zero, y-values barely raise off the x-axis, creating a bowl at the bottom of the parabola.

Now we create a short addition chain on the right side of the equation for two new quadratic equations.

$$y = x^2 + 1 \hspace{4cm} y = x^2 - 2$$

We find that the first equation shifts the graph up one unit and the second equation shifts the graph down two units.

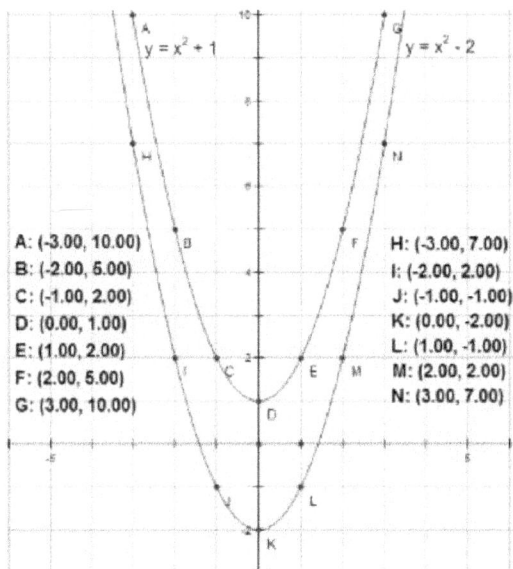

A: (-3.00, 10.00)
B: (-2.00, 5.00)
C: (-1.00, 2.00)
D: (0.00, 1.00)
E: (1.00, 2.00)
F: (2.00, 5.00)
G: (3.00, 10.00)

H: (-3.00, 7.00)
I: (-2.00, 2.00)
J: (-1.00, -1.00)
K: (0.00, -2.00)
L: (1.00, -1.00)
M: (2.00, 2.00)
N: (3.00, 7.00)

A constant added to a quadratic equation serves the dual purpose of acting as the y-intercept of the equation and, as a result, creating a vertical shift in the graph, up for positive, down for negative. We have seen this before in linear equations.

We increase the addition chains on the right side of the first equation with one more unlike term to create two new equations.

$$y = x^2 + 2x + 1 \hspace{3cm} y = x^2 - 2x + 1$$

190

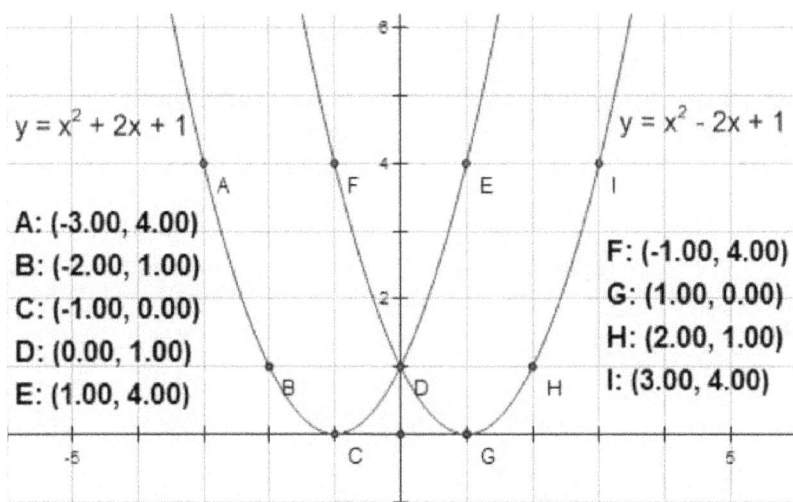

$y = x^2 + 2x + 1$ $y = x^2 - 2x + 1$

A: (-3.00, 4.00)
B: (-2.00, 1.00)
C: (-1.00, 0.00)
D: (0.00, 1.00)
E: (1.00, 4.00)

F: (-1.00, 4.00)
G: (1.00, 0.00)
H: (2.00, 1.00)
I: (3.00, 4.00)

Notice that, while the y-intercept is still + 1, the + 2x and – 2x terms cause horizontal shifts in the graph, but in the directions opposite to what one would expect, that is, left for + 2x, right for – 2x. This brings us to the general form of a quadratic equation.

$$y = ax^2 + bx + c$$

a, b and c are constants and a $\neq$ 0. Like the constants in a linear equation, the constants of a quadratic equation determine the direction and position of the parabola in the plane.

The constant c acts as the y-intercept of the parabola and thus causes vertical shifts up or down the y-axis. The coefficient b causes a horizontal shift in the graph. As will be seen shortly, the a and b coefficients determine the position of the vertex, or turning point, of the parabola. Thus, a, b and c all affect the position of the parabola in the plane.

The coefficient of the x^2 term, a, also determines the direction and width of the parabola. The positive sign on a results in a parabola which opens up; a negative a term gives a parabola which opens down. As the absolute value of a increases, the parabola becomes narrower, whether right side up for positive a or upside down for negative a.

191

$y = -x^2 - 2x + 3$

$y = -x^2 - 2x - 1$

$y = -2x^2 - 2x + 4$
$y = -x^2 - 2x + 3$

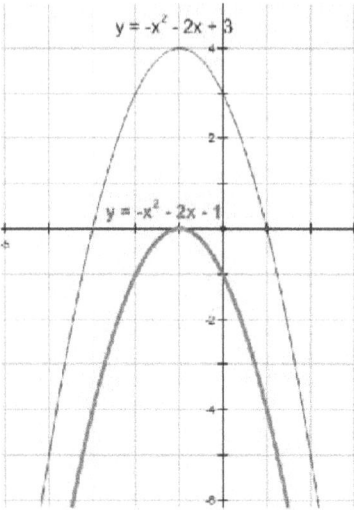

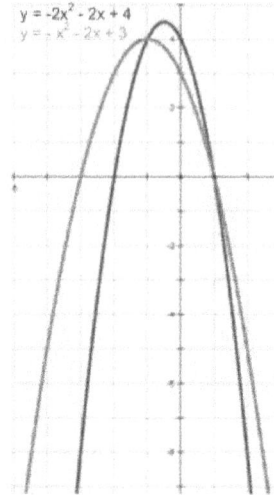

Now notice that, when a is negative, the b coefficient causes a horizontal shift in the expected direction, that is, left for $-2x$. This makes sense upon analysis of the formula for determining the x-coordinate of the vertex, or turning point, of the parabola.

$$x_{vertex} = \frac{-b}{2a}$$

If the x-coordinate of the vertex is positive, and a is positive, then negating negative b causes a positive, right shift consistent with a positive x-coordinate. If the x-coordinate of the vertex is negative, and a is positive, then negating positive b produces a consistent shift left of zero.

As with linear equations, there is more than one way to produce an accurate graph of a quadratic equation. Producing a graph with the greatest accuracy involves finding the x-coordinate of the vertex first, and choosing three or four integral x-values left and right of the vertex for substitution into the equation to find their corresponding y-values. So, given

$y = x^2 + 5x + 6$

We find the x-coordinate of the vertex first.

$$x_{vertex} = \frac{-b}{2a} = \frac{-(5)}{2(1)} = \frac{-5}{2}$$

Substituting this into the equation, we obtain the y-coordinate.

$$y = x^2 + 5x + 6 = (-5/2)^2 + 5(-5/2) + 6 = \frac{25}{4} - \frac{25}{2} + 6 = \frac{25}{4} - \frac{50}{4} + \frac{24}{4} = \frac{-1}{4}$$

vertex $(-5/2, -\frac{1}{4})$

We place this in a table and choose two integral values of x left and right of $-5/2$ for substitution into the equation to obtain y-coordinates. We also include x = 0 which gives us the y-intercept.

x	−4	−3	$-5/2$	−2	−1	0
y	2	0	−¼	0	2	5

We plot the points and connect them with a smooth curve. Note how the axes are scaled in order to accommodate the fractional values of the vertex's coordinates.

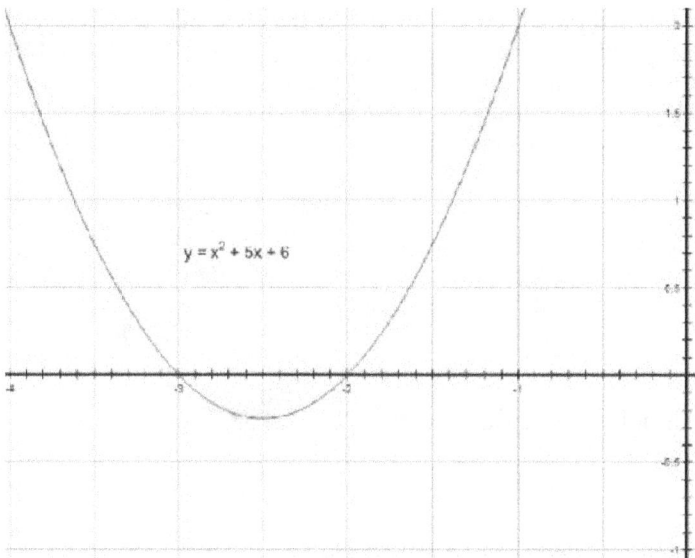

$y = x^2 + 5x + 6$

Interestingly, the integral x-values immediately left and right of the vertex in this equation happen to be the *x-intercepts* of the parabola. These x-values are interchangeably known as the *roots, solutions* or *zeros* of the equation.

As in the previous example, the most accurate method for sketching a parabola is to use a table of values where various x-coordinates are evaluated to find their corresponding y-coordinates for plotting the points. A commonly acceptable shortcut to graphing a parabola is to plot the coordinates of the vertex, x-intercepts and y-intercept. The vertex is found as before, and the y-intercept is found by setting $x = 0$, causing the $ax^2 + bx$ terms to vanish and leaving only c. Finding the *x-intercepts, also called roots, solutions or zeros*, of a quadratic equation has been visited on a number of occasions in this text: we set the equation equal to zero, because y equals zero for all points on the x-axis, and factor the equation to find the roots.

$$x^2 + 5x + 6 = 0$$
$$(x + 3)(x + 2) = 0$$

$$x = -3 \,|\, x = -2$$

We plot $(0, 5)$, $(-5/2, -1/4)$, $(-2, 0)$ and $(-3, 0)$ and connect the points with a smooth curve, producing the same graph as above.

Given how all quadratic equations produce the graph of a parabola and only a parabola, whether narrow or wide, opening up or opening down, it becomes clear from the graphs why three unlike terms in a second-degree addition chain form a unity that cannot be solved by unlinking the addition or subtraction links, but require factoring.

Find the coordinates of the x-intercepts, y-intercept and vertex in the following problems.

1. $x^2 - 10x + 24$
2. $x^2 - 3x + 2$
3. $x^2 - 3x - 4$

Graph the following equations.
4. $2x^2 + 6x + 4$
5. $-x^2 + 4$

Completing the Square

Regardless of the method used to sketch a parabola, finding the roots of the quadratic equation is always essential; the reason for this shall become evident shortly. Not all quadratic equations are easily factorable. This usually arises when the roots of the equation are not rational, that is, integral or fractional, but instead are *irrational*. In such instances, the technique of completing the square is a fast way to find the precise roots of the equation without using a graphing calculator.

Consider the equation

$$x^2 - 6x + 6 = 0$$

The equation cannot be factored to give rational roots because, as we shall see, the roots are irrational. Completing the square is the process of creating a perfect square trinomial and using it to solve for the roots.

$x^2 - 6x + 6 = 0$ — Subtract the c term, here, 6, from both sides of the equation.

$x^2 - 6x = -6$ — Take half of the b coefficient, here, -6, and square it. Add this term to both sides of the equation.

$$(-3)^2 = 9$$

$x^2 - 6x + 9 = 9 - 6$ — The left side is a perfect square trinomial; rewrite it as a binomial squared. The constant monomial of the binomial is half of b before squaring it. Combine like terms on the right side.

$(x - 3)^2 = 3$ — Take the square root of both sides of the equation.

$\sqrt{(x - 3)^2} = \pm\sqrt{3}$ — The square root of any term squared is the term. Both positive and negative radical three must be considered because, in

$x - 3 = \pm\sqrt{3}$ — reverse, each squared produces three. To isolate x, add three to both sides.

$x = 3 \pm \sqrt{3}$ — The two roots are three plus radical three and three minus radical three.

Evaluated on a calculator, $3 + \sqrt{3} \approx 4.732$ and $3 - \sqrt{3} \approx 1.268$. To sketch the graph of this equation, it would be better to find the coordinates of the vertex and substitute integral values of x into the equation for their y-values in order to obtain points which can be plotted.

195

To check if your solution is correct, apply these are two formulas which are easy to memorize:

Sum of roots = $\frac{-b}{a}$ Product of roots = $\frac{c}{a}$

$3 + \sqrt{3}$ $(3 + \sqrt{3})(3 - \sqrt{3}) = 9 - 3\sqrt{3} + 3\sqrt{3} - 3 = 6$
$\underline{+\ 3 - \sqrt{3}}$ add to zero
6 Therefore, b = - 6   (a = invisible 1)   and c = 6

Here, the mystery of roots' importance is revealed: like the roots of a plant or tree which give it life and character, the roots of a quadratic equation define the very equation itself and determine almost all of its characteristics.

As the next example illustrates, the technique of completing the square works equally well for finding integral roots, but is more tedious than factoring.

$x^2 + 6x + 8 = 0$ Subtract the c term, here, 8, from both sides of the equation.
$x^2 + 6x = -8$ Take half of the b coefficient, here, 6, and square it. Add to
$x^2 + 6x + 9 = 9 - 8$ both sides. Rewrite the left side as a binomial squared;
$(x + 3)^2 = 1$ combine like terms on the right. Take the square root of both
$\sqrt{(x + 3)^2} = \sqrt{1}$ sides; remember to take the positive and negative root of the
$x + 3 = \pm 1$ constant on the right.
$x = -3 + 1$ and $x = -3 - 1$
$x = -2$ and $x = -4$

Check: $-2 - 4 = -6$ and $(-2)(-4) = 8$
Therefore, b = 6 and c = 8

Complete the square to find the roots of the following equations. Check your answer by using the sum and product formulas on the roots you find. Remember that the sum of the roots equals $-b/a$ and the product of the roots equals c/a.

1. $y = x^2 - 4x + 3$
2. $y = x^2 - 4x - 1$
3. $y = x^2 - 8x + 14$
4. $y = x^2 - 10x + 20$
5. $y = x^2 - 7x + 12$

The Quadratic Formula

Occasionally, neither factoring nor completing the square lend themselves well to finding the roots of some quadratic equations. At such times, the quadratic formula is the most efficient way to find the roots of the equation.

$$x_1, x_2 = \frac{-b \pm \sqrt{b^2 - 4ac}}{2a}$$

Consider the graphs of the quadratic equations below:

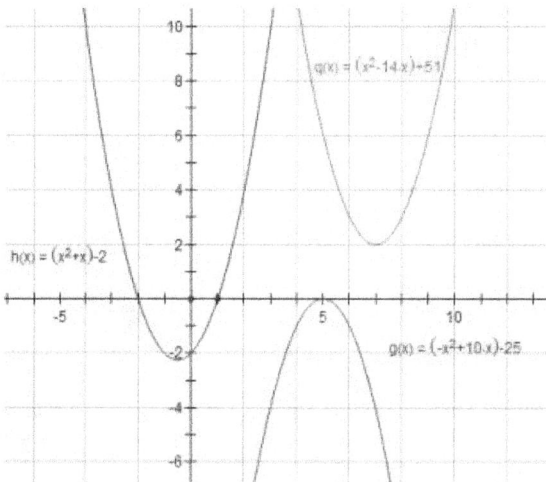

The first parabola on the left intersects the x-axis in two distinct points. The second parabola is hanging upside down from its vertex on the x-axis; the equation for this parabola has a single root at the vertex. The third parabola on the right does not intersect the x-axis at all and has what are called imaginary roots, studied in greater detail in more advanced algebra courses. The equation for this third parabola requires the use of the quadratic formula for finding the roots of its equation.

For our purposes, the quadratic formula is useful when one or both roots are rational, that is, fractional, as in the next example.

197

$2x^2 - 3x + 1 = 0$

We begin by identifying a, b and c from the equation.

$a = 2$ $\qquad\qquad$ $b = -3$ $\qquad\qquad$ $c = 1$

We make the appropriate substitutions into the quadratic formula.

$$x_1, x_2 = \frac{-b \pm \sqrt{b^2 - 4ac}}{2a}$$

$$x_1, x_2 = \frac{-(-3) \pm \sqrt{(-3)^2 - 4(2)(1)}}{2(2)}$$

$$x_1, x_2 = \frac{3 \pm \sqrt{9 - 8}}{4}$$

$$x_1, x_2 = \frac{3 \pm \sqrt{1}}{4}$$

$$x_1, x_2 = \frac{3 \pm 1}{4}$$

$$x_1 = \frac{3 + 1}{4}, \qquad x_2 = \frac{3 - 1}{4}$$

$$x_1 = \frac{4}{4} = 1, \qquad x_2 = \frac{2}{4} = \frac{1}{2}$$

To check your solutions, add and multiply the roots.

$$1 + \frac{1}{2} = \frac{3}{2}, \quad \frac{-b}{a} = \frac{3}{2}, \quad b = -3, a = 2, \quad (1)(\tfrac{1}{2}) = \frac{1}{2} = \frac{c}{a}, \quad c = 1$$

The quadratic formula can be used to find all roots of any nature: real and integral; real and rational; real and irrational; and imaginary. The next example applies the quadratic formula to finding irrational roots.

198

$$2x^2 - 6x + 2 = 0$$

$$a = 2 \qquad\qquad b = -6 \qquad\qquad c = 2$$

$$x_1, x_2 = \frac{-b \pm \sqrt{b^2 - 4ac}}{2a}$$

$$x_1, x_2 = \frac{-(-6) \pm \sqrt{(-6)^2 - 4(2)(2)}}{2(2)}$$

$$x_1, x_2 = \frac{6 \pm \sqrt{36 - 16}}{4}$$

$$x_1, x_2 = \frac{6 \pm \sqrt{20}}{4}$$

$$x_1, x_2 = \frac{6 \pm 2\sqrt{5}}{4} = \frac{3}{2} \pm \frac{\sqrt{5}}{2}$$

$$x_1 = \frac{3}{2} + \frac{\sqrt{5}}{2}, \qquad x_2 = \frac{3}{2} - \frac{\sqrt{5}}{2}$$

To check your solutions, add and multiply them and compare to the original equation.

$$\frac{3}{2} + \frac{\sqrt{5}}{2} \qquad\qquad (\frac{3}{2} + \frac{\sqrt{5}}{2})(\frac{3}{2} - \frac{\sqrt{5}}{2}) = \frac{9}{4} - \frac{5}{4} = \frac{4}{4} = \frac{c}{a} \, but \; a = 2 \; so \; c = 2$$

$$+ \frac{3}{2} - \frac{\sqrt{5}}{2}$$

$$\frac{6}{2} = \frac{-b}{a} \quad \therefore \; b = -6 \; and \; a = 2$$

One might question how this equation is different from $x^2 - 3x + 1 = 0$; after all, when a factor of two is removed from all terms of $2x^2 - 6x + 2 = 0$,

199

the roots are identical for both equations. The answer can be seen in the graph below:

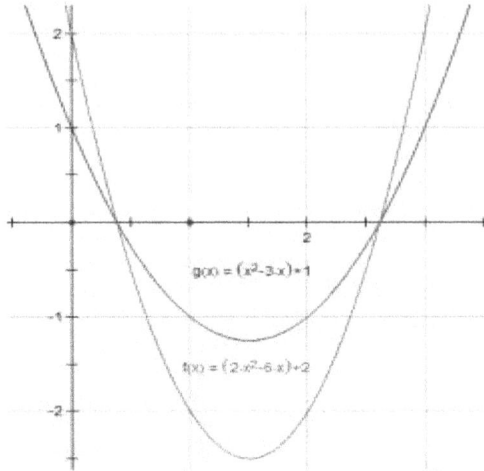

While both equations and their graphs share the same roots, the factor of 2 narrows the parabola for the equation $2x^2 - 6x + 2 = 0$ and gives it a y-intercept of 2 instead of 1. Unlike linear equations, where multiples of the same equation create an identical graph of a line, this is not the case with quadratics. Multiples of quadratic equations produce parabolas with identical roots but otherwise different characteristics.

Apply the quadratic formula to the following problems.   Check your solutions using the sum and product formulas to see if the original equations are produced.

1. $2x^2 - x - {}^3/_8 = 0$
2. $2x^2 - 8x - 6 = 0$
3. $3x^2 - 6x - 3 = 0$
4. $x^2 - 2x - 2 = 0$
5. $x^2 - 4x + 1 = 0$

Appendix A:

Summary Chart
of
Operations of Algebra

	Integers	Fractions	Decimals	Scientific Notation	Monomials
When are They Like terms?	.ALWAYS	When they have COMMON DENOMINATORS	ALWAYS.	When the bases 10 have the same exponent; then, the powers are like terms.	When they have: 1. The same variable base; AND 2. The same exponent on that base.
Addition	# 1: **SAME SIGNS? ADD.** $-7 + -4 = -11$ # 2: **DIFFERENT SIGNS? SUBTRACT.** $-7 + 4 = -3$ $-4 + 7 = 3$	Give fractions COMMON DENOMINATORS. **ADD** numerators using the two rules at left; **KEEP** com. denominator. $\frac{1}{2} + \frac{3}{5} = \frac{5+6}{10}$ $= \frac{11}{10}$ $\frac{1}{2} + \frac{-3}{5} = \frac{5-6}{10} = \frac{-1}{10}$	Line up decimals; **ADD** columns from right to left. 2.34 + - 10.04 $- \mid 10.04 - 2.34 \mid$ $= - 7.7$	**ADD** coefficients – **KEEP** common base ten – **KEEP** common exponent. Adjust the decimal in the coefficient and exponent in advance, if powers are unlike.	**ADD** coefficients – **KEEP** common base – **KEEP** common exponent. Link unlike terms in an addition chain. $2x^2 + -4x^2 = -2x^2$
Subtraction	ADD THE **OPPOSITE: KEEP** sign of the first term – change the minus sign to a **PLUS** sign – change the negative when following this rule. Once the subtraction operation has been changed into addition, follow the SUBTRACTION OPERATION can be changed to ADDITION by following the procedure **KEEP – PLUS ADDITIVE INVERSE, SUBTRACTION.** Unlike terms are then linked together in an addition chain.				
Multiplication	# 1: **SAME** signs ? **MULTIPLY** factors to obtain a **POSITIVE** product. # 2: **DIFFERENT** signs? **MULTIPLY** to obtain a **NEGATIVE** product.	**MULTIPLY** numerators to obtain product numerator. **MULTIPLY** denominators to obtain a product denominator. Reduce to lowest terms. $\frac{2}{5} \cdot \frac{-3}{4} = \frac{-6}{20} = \frac{-3}{10}$	**MULTIPLY** as if the decimals were not present; the PRODUCT should have as many decimal places as are in the sum of the factors.	**MULTIPLY** coefficients – **KEEP** base ten – **ADD** exponents. $2 \times 10^2 \times 6.02 \times 10^{23}$ $= 12.04 \times 10^{25}$	**MULTIPLY** coefficients – **KEEP** like bases – **ADD** exponents of like bases. $3x^2 \cdot 4x^5 = 12x^7$
Division	# 1: **SAME** signs on dividend and divisor yield a **POSITIVE** quotient. # 2: **DIFFERENT** signs on dividend and divisor yield a **NEGATIVE** quotient.	**MULTIPLY** first fraction by RECIPROCAL of second fraction; use multiplication rules above. $\frac{2}{3} \div \frac{-4}{3} = \frac{2}{3} \cdot \frac{-3}{4} = $ $= \frac{-1}{2}$	Move decimal to the right completely out of divisor; move decimal same number of times right in dividend; place decimal directly above in quotient.	**DIVIDE** coefficients – **KEEP** base ten – **SUBTRACT** exponents. $\frac{12.04 \times 10^{25}}{2 \times 10^2}$ $= 6.02 \times 10^{23}$	**DIVIDE** coefficients - **KEEP** like bases – **SUBTRACT** exponents. $\frac{12x^7}{3x^2} = 4x^5$

Binomials	Polynomials	Radicals	Algebraic Fractions
A binomial is comprised of two monomials, usually unlike, linked by addition or subtraction. A polynomial is three or more unlike monomials so linked. Rarely, they are like terms when their MONOMIALS AND CONNECTORS ARE IDENTICAL LINK-FOR-LINK. In most instances, binomials and polynomials are unlike terms linked in an addition chain.		When they have the SAME RADICAND. If radicands are unlike, simplify them to have the smallest possible radicand; then combine like radicals.	When they have the SAME monomial/binomial/polynomial DENOMINATOR.
Place like monomials in columns and **ADD**, using the two rules at far left: $$\begin{array}{rrr} x^2 & x & \#s \\ 3x^2 & -x & -4 \\ -5x^2 & 8x & -1 \\ \hline -2x^2 + & 7x + & -5 \\ = -2x^2 + & 7x - & 5 \end{array}$$ Main idea: **ADD – KEEP – KEEP**. Link unlike terms in an addition chain.		Give radicals LIKE RADICANDS by factoring. Then, **ADD** coefficients – **KEEP** like radical. Link unlike terms in an addition chain. $$\sqrt{72} + \sqrt{50}$$ $$= \sqrt{36}\cdot\sqrt{2} + \sqrt{25}\cdot\sqrt{2}$$ $$= 6\sqrt{2} + 5\sqrt{2} = 11\sqrt{2}$$	Give fractions COMMON DENOMINATORS. **ADD** numerators using all applicable addition rules at left – **KEEP** common denominator. Combine all like terms in numerator; link unlike terms in an addition chain over denominator. $$\frac{2x}{3} + \frac{3}{4x^2} = \frac{8x^3 + 9}{12x^2}$$

sign of the second term to its **OPPOSITE**. Unsigned terms are positive and should be changed to addition rules above. As with addition, only LIKE TERMS can be combined by subtraction. EVERY **OPPOSITE**.   NOTE: LIKE TERMS MAY BE COMBINED ONLY BY MEANS OF ADDITION AND ITS

Binomials	Polynomials	Radicals	Algebraic Fractions
FOIL: Multiply First monomials– Outer monomials – Inner monomials – Last monomials, using **MULTIPLY – KEEP – ADD** rules at left; combine like terms; link unlike terms in an addition chain.	Apply the distributive property as many times as in the shorter addition chain, using **MULTIPLY – KEEP– ADD**. Combine like terms; link unlike terms in an addition chain.	**MULTIPLY** coefficients – **MULTIPLY** radicands. Simplify new radicand by factoring. $$2\sqrt{18} \cdot 4\sqrt{2} = 8\sqrt{36}$$ $$= 8 \cdot 6 = 48$$ $$6\sqrt{7} \cdot 2\sqrt{14} = 12\sqrt{98}$$ $$= 12\sqrt{49}\sqrt{2} = 12 \cdot 7\sqrt{2}$$ $$= 84\sqrt{2}$$	Reduce common factors of monomials and binomials in numerators and denominator. Then **MULTIPLY** numerators –**MULTIPLY** denominators. Use all applicable rules at left. $$\frac{-24}{x} \cdot \frac{x+3}{2x+6} = \frac{-24}{x} \cdot \frac{x+3}{2(x+3)}$$ $$= \frac{-12}{x}$$
Pull the greatest common factor from a binomial and set it in front of parentheses around the reduced binomial. Factor the remaining binomial when possible. $$4x^2 - 64 = 4(x^2 - 16)$$ $$= 4(x+4)(x-4)$$	Factor a quadratic trinomial into two binomials by guessing the roots whose sum and product are the - b and +c constants. $$x^2 - 8x + 12$$ $$= (x-6)(x-2)$$ $$-6 + -2 = -8$$ $$(-6)(-2) = 12$$	**DIVIDE** coefficients – **DIVIDE** radicands. Simplify new radicands by factoring. $$\frac{12\sqrt{98}}{3\sqrt{2}} = \frac{12}{3}\sqrt{\frac{98}{2}}$$ $$= 4\sqrt{49} = 4 \cdot 7 = 28$$	**MULTIPLY** first fraction by RECIPROCAL of second fraction; use the multiplication rules above. $$\frac{5x + 25}{4x - 8} \div \frac{6x + 30}{2}$$ $$= \frac{5x + 25}{4x + 8} \cdot \frac{2}{6x + 30}$$ $$= \frac{5(x + 5)}{4(x - 2)} \cdot \frac{2}{6(x + 5)}$$ $$= \frac{5}{12(x - 2)}$$

203

Appendix B

Decimal Place Value
and
Scientific Notation

10^4	10^3	10^2	10^1	10^0		$\frac{1}{10^1}$	$\frac{1}{10^2}$	$\frac{1}{10^3}$	$\frac{1}{10^4}$
						10^{-1}	10^{-2}	10^{-3}	10^{-4}
5	9,	2	6	4 .		7	8	3	1

$$= (50,000) + (9,000) + (200) + (60) + (4) + (^7/_{10}) + (^8/_{100}) + (^3/_{1,000}) + (^1/_{10,000})$$

$$= (5 \times 10,000) + (9 \times 1,000) + (2 \times 100) + (6 \times 10) + (4 \times 1) + (7 \times .1) + (8 \times .01) + (3 \times .001) + (1 \times .0001)$$

$$= (5 \times 10^4) + (9 \times 10^3) + (2 \times 10^2) + (6 \times 10^1) + (4 \times 10^0) + (7 \times 10^{-1}) + (8 \times 10^{-2}) + (3 \times 10^{-3}) + (1 \times 10^{-4})$$

The number 59,264.7831 (read, "Fifty-nine thousand, two hundred sixty-four and seven-tenths and eight-one-hundredths and three-one-thousandths and one-ten-thousandth") can be expressed in standard form, or as a sum of place values written in standard notation, or as a sum of products indicating the place value of each digit, or as a sum of products in scientific notation.

Note that *__any__* real number raised to the zero power will always have a value of ONE (1). Mathematically, this is expressed as follows: $a^0 = 1$, $a \in \mathcal{R}$ ("*a* to the zero power equals one, where *a* is an element of the set of real numbers.").

ADDITION

Adding Integers - p. 7

1. -2	2. 2	3. -10	4. -4
5. 4	6. -10	7. -13	8. 3
9. -3	10. -7	11. 7	12. -11
13. -7	14. 7	15. -15	16. -12
17. -8	18. 8		

Adding Numerical Fractions - pgs. 12 - 14
1. 23/20 2. $-7/20$ 3. 1/12 4. $-5/12$ 5. $-7/10$ 6. $-37/20$
7. 17/4 8. 19/15 9. $-69/54$ 10. 10/3

Adding Decimals - p. 16
1. 13.75096 2. 379.83777 3. 549,578.6288 4. -255.59
5. $-26,957.22$ 6. 11,954,136,395.1844 7. $-12,046,004,568.875$

Adding Numbers Expressed in Scientific Notation - p. 18 - 19
1. -8.2109×10^4 2. -2.443371×10^8 3. -1.2004074×10^3 4. 2.77846×10^1
5. -1.09659763×10^5 6. -4.8608688×10^2 7. -4.48188×10^3

Adding Monomials - p. 24

1. $-2x^2$ 2. $-6c^2d$ 3. $3xy^2$ 4. $-8d$ 5. $12xy^2$ 6. $.16m^2n$
7. $-.35a^2b$ 8. $-1\,^1/_{24}\,x^2y$ 9. $^1/_{24}\,xy^2$ 10. $^1/_{24}\,r^2s$ 11. $-x+10$
12. $-1.5a+4$ 13. $-2x+-2$ 14. $-1/6\,xy^2+11/12\,x^2y$
15. $^{-11}/_{12}\,a^2b+\,^{13}/_{12}\,ab^2$ 16. x^2+5x+6 17. x^2+3x+2
18. $.1y^2+.6y+.8$ 19. $^{-5}/_{24}\,x^2+\,^7/_3\,x+2$ 20. $^3/_4\,y^2+\,^{-5}/_8\,y+-3$

Adding Binomials and Polynomials - p. 27

1. $9x^2+-8$ 2. $9x^2+8$ 3. $7x^2+-3$ 4. $-3x^2+2x+1$
5. $2x^2+-9x+6$ 6. $-10x^2+-4x+-2$ 7. $5x^2+5x+2$
8. $-4x^2+-11x+3$ 9. $20x^2+42xy+25y^2$
10. $2x^3+-6x^2y+-6xy^2+-26y^3$ 11. $2x^4+-4x^3y+30x^2y^2+-28xy^3+17y^4$

Adding Radicals - p. 32, 34

1. $-3\sqrt2$ 2. $6\sqrt3$ 3. $\sqrt5$ 4. $8x\sqrt{3x}$ 5. $-12x^2\sqrt{x}$
6. $-3\sqrt5$
7. $-5\sqrt{2x}+-2x\sqrt{2x}$
8. $-10x\sqrt2+-2x\sqrt3+3\sqrt2$
9. $-26x\sqrt{3x}+7\sqrt{3x}+7\sqrt{2x}$

10. $x^2\sqrt{5x} + -3x\sqrt{2x} + -23x\sqrt{5x}$

11. $7\sqrt{2}$　　12. $45\sqrt{2}$　　　13. $102\sqrt{3}$　　　14. $202\sqrt{3}$　　15. $65\sqrt{5}$
16. $44\sqrt{6}$　　17. $6x\sqrt{2} + 26\sqrt{2}$　　18. $2\sqrt{7} + -5\sqrt{6}$　　19. $-2\sqrt{6} + 4x\sqrt{6}$
20. $2\sqrt{6} + 2\sqrt{5}$

Adding Algebraic Fractions - pgs. 38 – 39, 42 - 43

1. $\dfrac{23x}{20}$　　2. $\dfrac{-x}{x+4}$　　3. $\dfrac{2}{24x}$ or $\dfrac{1}{12x}$　　4. $\dfrac{-13}{40x}$　　5. $\dfrac{-2x}{x^2-5x+6}$

6. $\dfrac{-12x-25}{20x^2}$　　7. $\dfrac{-29}{12x}$　　8. $\dfrac{40x-21}{15x^2}$　　9. $\dfrac{-8x^2-15}{18x^2}$

10. $\dfrac{4+6x}{3x^2}$　　11. $\dfrac{-7x+15}{x(x-3)}$　　12. $\dfrac{2x+-82}{(x+8)(x-6)}$　　13. $\dfrac{-2x+-40}{x(x+5)}$

14. $\dfrac{3x+-22}{x(x+11)}$　　15. $\dfrac{-x+-56}{x(x-7)}$　　16. $\dfrac{65x^2+-4x+20}{5x^2(x-5)}$　　17. $\dfrac{5x+-13}{(x+1)(x+7)}$

18. $\dfrac{3x+-22}{x(x+11)}$　　19. $\dfrac{-x+-56}{3x(x-7)}$　　20. $\dfrac{3x^2+-4x+19}{(x-4)(x^2+1)}$

SUBTRACTION

Subtracting Integers - p. 48

1. 10　　　2. 2　　　　3. -2　　4. -10　　5. 10
6. -10　　7. 4　　　　8. -3　　9. 13　　　10. -13
11. 11　　12. -11　　13. -7　　14. 7　　　15. -15
16. 15　　17. -7　　18. -8　　19. -12　20. 12

Subtracting Numerical Fractions – pgs. 49 - 51
1. 7/20　　2. $-23/20$ or $-1\,^3/_{20}$　　　3. 5/12　　4. $-1/12$　5. $-3/10$
6. 13/20　7. 19/4 or $4\,^3/_4$　　8. 61/15 or $4\,^1/_{15}$　　9. 21/54 or 7/18　10. $-2/3$

Subtracting Decimals - p. 52
1. 13.75096　　　2. 379.83777　　　3. 549,578.6288　　　4. -255.59
5. $-26,957.22$　　6. 11,954,136,395.1844　　7. $-457,235,009.998547$
8. 178,930,926.0085759334

Subtracting Numbers Expressed in Scientific Notation - p. 53

1. 3.742029×10^8　　　2. -1.2004074×10^3　　3. 2.39094×10^1　　4. -1.09070237×10^5
5. -5.0149312×10^2　　6. 3.18712×10^3　　7. -2.05526165×10^5
8. 1.447693176×10^6　　9. 7.960465813×10^8

Subtracting Monomials - pgs. 54 - 55

1. $-12x^2$ 2. $-2c^2d$ 3. $-9xy^2$ 4. $-24d$ 5. $-12xy^2$ 6. $1.04m^2n$

7. $-.11a^2b$ 8. $43/24\ x^2y$ or $1\ ^{19}/_{24}\ x^2y$ 9. $^5/_{24}\ xy^2$ 10. $^{-19}/_{24}\ r^2s$

11. $3x-10$ 12. $2.5a-4$ 13. $20x+2$

14. $-5/6\ xy^2 - 11/12\ x^2y$ 15. $^5/_{12}\ a^2b - ^{13}/_{12}\ ab^2$

16. x^2-5x-6 17. $3\ x^2-3x+2$ 18. $.5y^2-.6y-.8$

19. $^{37}/_{24}\ x^2 - ^7/_3\ x - 2$ 20. $^5/_4\ y^2 + ^5/_8\ y + 3$

Subtracting Binomials and Polynomials – p. 56

1. $-9x^2+2$ 2. x^2+5 3. $-7x^2+4x-17$ 4. $22x^2+x+2$

5. $-4x^2-14x$ 6. $15x^2+17x$ 7. $8x^2-3x+7$ 8. $-12x^2-22xy-7y^2$

9. $12x^2y+12xy^2+28y^3$ 10. $12x^3y+30x^2y^2+36xy^3-15y^4$

Subtracting Radicals - pgs. 57 - 58

1. $-7\sqrt{2}$ 2. $4\sqrt{3}$ 3. $\sqrt{5}$ 4. $-2x\sqrt{3x}$

5. $20\ x^2\sqrt{x}$ 6. $13\sqrt{5}$

7. $\sqrt{2x}+2x\sqrt{2x}$

8. $-2x\sqrt{2}+2x\sqrt{3}-3\sqrt{2}$

9. $8x\sqrt{3x}-7\sqrt{3x}-7\sqrt{2x}$

10. $15\ x^2\sqrt{5x}+3x\sqrt{2x}+23\ x\sqrt{5x}$

11. $-3\sqrt{2}$ 12. $-3\sqrt{2}$ 13. $-16\sqrt{3}$ 14. $-26\sqrt{3}$

15. $-9\sqrt{5}$ 16. $-3\sqrt{6}$ 17. $42x\sqrt{2}-13\sqrt{2}$ 18. $8\sqrt{7}+5\sqrt{6}$

19. $-2\sqrt{6}-4x\sqrt{6}-2\sqrt{5}$ 20. $-6\sqrt{6}-2\sqrt{5}$

Subtracting Algebraic Fractions - pgs. 58 - 62

1. $\dfrac{7x}{20}$ 2. $\dfrac{-5x}{x+4}$ 3. $\dfrac{10}{24x}$ or $\dfrac{5}{12x}$ 4. $\dfrac{-3}{40x}$ 5. $\dfrac{x}{x^2-5x+6}$

6. $\dfrac{-12x+25}{20x^2}$ 7. $\dfrac{13}{12x}$ 8. $\dfrac{40x+21}{15x^2}$ 9. $\dfrac{-24x^2+45}{54x^2}$ 10. $\dfrac{4-6x}{3x^2}$

11. $\dfrac{-8x+15}{x(2x-3)}$ 12. $\dfrac{12x-2}{(x-6)(x+8)}$ 13. $\dfrac{14x+40}{x(x+5)}$ 14. $\dfrac{7x+22}{x(x+11)}$

15. $\dfrac{-11x+64}{(x-6)(x-7)}$ 16. $\dfrac{-12x+36}{(x-2)(x-5)}$ 17. $\dfrac{-11x+4}{(x+1)(x-4)}$ 18. $\dfrac{-10x+68}{(x-6)(x-7)}$

19. $\dfrac{-13x+41}{(x-2)(x-5)}$ 20. $\dfrac{-3x^3-7x+16}{(x^2+1)(x-4)}$

MULTIPLICATION
Multiplying Integers - p. 69

1. -24 2. 24   3. 24   4. -24 5. -21 6. -21 7. 21
8. 40   9. -40 10. -40 11. -18 12. -18 13. 18   14. 18
15. -44 16. -44 17. 44   18. 20

Multiplying Numerical Fractions - pgs. 69 - 71
1. 10/9   2. $-35/48$ 3. $-64/15$ 4. 18/115
5. $-15/56$ 6. $-2/108$ or $-1/54$ 7. 16/1898. $-20/3$ or $-6^2/_3$
9. -12 10. 6/121   11. $-1/10$ 12. $-1/27$ 13. 40/9

Multiplying Decimals - p. 72
1. $-3,635,324.471$ 2. $-3,613,358.408$ 3. $-6,892.46487$

Multiplying Numbers Expressed in Scientific Notation - p. 74
1. $1.087298955 \times 10^{11}$ 2. $1.087298955 \times 10^{-1}$ 3. $4.894730662 \times 10^{-8}$
4. $9.389031591 \times 10^{9}$ 5. $-2.212316877 \times 10^{9}$ 6. -4.40617835×10
7. $3.157966956 \times 10^{-1}$ 8. 1.533105×10^{8} 9. $2.007168847 \times 10^{11}$
10. $7.165655217 \times 10^{-5}$

Multiplying Monomials - pgs. 77 - 78
1. $-8x^6$ 2. $-7x$ 3. $\dfrac{-40}{x^2}$ 4. $\dfrac{-90}{x^6}$ 5. $\dfrac{33}{x^3}$ 6. $-27x^6$ 7. $-504x^5z^3$
8. $-10ab^4c^4$ 9. $\dfrac{-48z^3}{x^8}$ 10. $\dfrac{p^4q^2}{r}$ 11. $25x^6 y^8 z^{10}(x + y + z)^2$
12. $\dfrac{-77h^7k^4 (j + k)^2}{j}$ 13. $\dfrac{j^{21}(h - j - k)^2}{54h^4k^2}$

Multiplying a Monomial by a Polynomial - p. 80
1. $-6x + 42$ 2. $5x^2 - 20x + 15$ 3. $-4x^2 + 8x - 4$
4. $4\sqrt{2}\, x^2 - 5\sqrt{2}\, x + \sqrt{2}$ 5. $5x^3y + 10x^2y^2 + 5xy^3$
6. $4x^4z - 32x^3yz + 96x^2y^2z - 128xy^3z + 64y^4z$

Multiplying Binomials - p. 83
1. $x^2 - 3x - 54$ 2. $x^2 + 9x + 20$ 3. $x^2 - 12x + 36$ 4. $x^2 - 4x - 32$
5. $8x^2 + 14x - 35$ 6. $x^2 - 16$

Multiplying Polynomials - pgs. 84 - 85
1. $x^4 - 4x^3 + 6x^2 - 8x + 8$ 2. $-20x^4 + 12x^3 - 37x^2 + 3x - 8$
3. $-x^4 + 10x^3 - 19x^2 + 16x + 64$ 4. $-2x^4 + 23x^3 + 35x^2 + 47x + 65$
5. $4x^4 + 5x^3 - 8x - 7$

Multiplying Radicals – pgs. 87 - 88

1. $12x$ 2. $16\sqrt{x}$ 3. $18x\sqrt{x}$ 4. $14x\sqrt{x}$ 5. $24\sqrt{xy}$

6. $24xy$ 7. $25\, xy\sqrt{xy}$ 8. 24   9. $60\sqrt{3}$

10. 24   11. 12   12. $12\sqrt{5}$ 13. 72   14. $12\sqrt{6}$

Multiplying Algebraic Fractions - pgs. 89 -- 91

1. $\dfrac{-3x}{10}$ 2. $\dfrac{-2}{x}$ 3. $\dfrac{12}{x}$ 4. $\dfrac{3}{2x}$ 5. $\dfrac{-3x}{8}$ 6. $\dfrac{x}{2}$

7. $\dfrac{-1}{7x^2}$ 8. $\dfrac{13x}{4}$ 9. $\dfrac{-x}{16}$ 10. $\dfrac{-5}{24}$ 11. $\dfrac{5}{2x(x-4)}$ 12. $\dfrac{-91}{2x}$

13. $\dfrac{-12}{x}$ 14. $\dfrac{2}{x-4}$ 15. $\dfrac{-2}{x(x+3)}$ 16. $\dfrac{1}{x}$ 17. 10   18. $x+3$

19. $\dfrac{8x^2}{(x+7)(x+5)}$ 20. $\dfrac{3(x-2)}{5(x+2)}$

DIVISION

Dividing Integers - pgs. 94 - 95

1. -16 2. 12   3. -4 4. 17   5. -14 6. -15 7. -5
8. -4 9. -18 10. 6   11. -14 12. -13 13. 16   14. 18
15. 17   16. 23

Dividing Numerical Fractions - pgs. 96 - 97
1. $-5/6$ 2. $-2/3$ 3. $2/3$ 4. 1   5. $13/5$ 6. $-1/3$ 7. $1/108$
8. $1/18$ 9. $-1/35$ 10. $-7/2$ 11. $-4/19$ 12. 63

Dividing Decimals – pgs. 98 -- 99
1. 2.472   2. 35.24   3. 432   4. 53.7   5. 44.5   6. 637.28

Dividing Numbers Expressed in Scientific Notation - p. 99
1. 8.0×10^1 2. 8.0×10^{13} 3. 6.3×10^{-2}
4. 6.3×10^{-5} 5. -2.65×10^2 6. -5.0×10^{-5} 7. 3.33×10^{-9}
8. 4.45×10^{-17} 9. -4.86×10^3 10. 1.325×10^2

Dividing Monomials - p. 103 - 104
1. $-4x^4$ 2. $\dfrac{4}{x^4}$ 3. $-7x$ 4. $\dfrac{1}{4x^4}$ 5. $\dfrac{-14}{x^3}$ 6. $\dfrac{-3.1}{x^{17}y}$

7. $\dfrac{-7}{x^2y^{10}}$ 8. $\dfrac{x^{13}y^9(x+y)^2}{3}$ 9. $\dfrac{-4x^5}{3y^2z^3}$ 10. $\dfrac{-(x+y)}{x^8\,y\,z^2}$

Factoring Binomials - pgs. 105 - 107
1. $14x(x-3)$ 2. $-5(5x-9)$ 3. $-13x(x-4)$ 4. $-18x(x+3)$
5. $-3x(x+15)$ 6. $x(x-1)$ 7. $17x(x+5)$ 8. $6x(x-10)$
9. $-12x(x-11)$ 10. $8x(x-8)$ 11. $(x+9)(x-9)$ 12. $(x+11)(x-11)$
13. $(2x+8)(2x-8)$ 14. $(3x+10)(3x-10)$ 15. $(4x+12)(4x-12)$
16. $(5x+13)(5x-13)$ 17. $(14+6x)(14-6x)$ 18. $(15+7x)(15x-7x)$
19. $(16+20y)(16-20y)$ 20. $(y/17+22x^2)(y/17-22x^2)$ 21. $3(x+3)(x-3)$
22. $4(x+5)(x-5)$ 23. $2(x+7)(x-7)$ 24. $5(x+6)(x-6)$
25. $6(x+10)(x-10)$ 26. $4(x+9)(x-9)$ 27. $5(x+11)(x-11)$
28. $3(x+13)(x-13)$ 29. $12(x+12)(x-12)$ 30. $16(x^2+16)(x+4)(x-4)$

213

Dividing a Binomial by a Monomial - p. 108
1. $5x^5 + 3x$ 2. $4x^8 - 2x$ 3. $x^2 - 16x$
4. $-3x^2 - 2$ 5. $8x - ^6/_7$ 6. $-1.8x^2y^2 - 2x$
7. $\dfrac{1}{2x} - x^4$ 8. $\dfrac{4}{x^4y} - \dfrac{3}{2x^5}$

Factoring Trinomials - pgs. 113 - 115
1. $(x+1)(x+5)$ 2. $(x+1)(x+8)$ 3. $(x+3)(x+4)$ 4. $(x-3)(x-4)$
5. $(x-4)(x+3)$ 6. $(x-2)(x-3)$ 7. $(x+4)(x-9)$ 8. $(x-12)(x-4)$
9. $(x-36)(x+4)$ 10. $(x-16)(x+4)$ 11. $(-x-4)(x+3)$ 12. $(-x-7)(x+3)$
13. $(-x+5)(x-11)$ 14. $(-x+6)(x-9)$ 15. $(2x-3)(x+7)$
16. $(2x+2)(x-5)$ 17. $(3x+2)(x-5)$ 18. $(3x+3)(x+6)$
19. $(x+8)^2$ 20. $(x-10)^2$ 21. $(x-21)^2$
22. $(x-23)^2$ 23. $(x+24)^2$ 24. $(x-14)^2$
25. $(x+19)^2$ 26. $(x+18)^2$ 27. $(x+22)^2$
28. $(x-25)^2$

Factoring and Dividing Radicals - p. 117
1. $1/4$ 2. $1/8$ 3. $1/3$ 4. 7 5. 8
6. $1/6$ 7. 5 8. 6 9. $x/16$ 10. $\dfrac{11}{\sqrt{x}}$ 11. $13\sqrt{x}$ 12. $6 + \sqrt{13}$

13. $\dfrac{5 - \sqrt{3}}{3}$ 14. $-8 - 4\sqrt{2}$

Factoring and Dividing Algebraic Fractions - p. 118

1. $\dfrac{x+2}{x-2}$ 2. $\dfrac{x+6}{x+2}$ 3. $\dfrac{x+1}{x+2}$ 4. $\dfrac{x-4}{x-3}$ 5. $\dfrac{x-6}{x+1}$

6. $\dfrac{(x+13)(x-1)}{(x-13)(x+1)}$ 7. $\dfrac{x+4}{x-5}$ 8. $\dfrac{x+1}{x-1}$ 9. $\dfrac{x-3}{x+3}$

ORDER OF OPERATIONS

Evaluating Numerical Expressions - pgs. 122 - 123
1. 19 2. -668 3. -126 4. 56 5. 784
6. 10 7. $-3,509$ 8. $-3,180$ 9. -600

Evaluating Algebraic Expressions - p. 123 - 124
1. 52 2. $-4/3$ 3. $-2/0$ undefined   4. 40
5. $1/64$ 6. -127 7. $4,096$ 8. $-1,290,240$

APPLICATIONS IN ONE VARIABLE: X AS AN UNKNOWN
Solving Single Variable First Degree Equations in One Step - p. 130
1. 16 2. 32 3. 3 4. 192 5. 6
6. 18 7. 2 8. 72 9. 5 10. 17
11. $11/6$ 12. 66 13. -25 14. 27 15. $1/26$
16. 26 17. -28 18. -22 19. $-25/3$
20. -75 21. -22 22. -30 23. 6
24. 104 25. 9 26. -23 27. $7/16$
28. 112 29. -32 30. 16 31. $-1/3$ 32. -192
33. 3.6 34. 1.8 35. -3 36. -2.43

Solving Single Variable First Degree Equations in Two Steps - p. 133
1. 4   2. 8   3. 16   4. 32
5. − 4   6. − 8   7. − 36   8. − 72
9. 7   10. 3.5   11. 112   12. 56   13. − 21/22
14. − 23/22   15. −132   16. 88   17. − 11/14 18. 5/14
19. − 154 20. 70   21. − 6   22. − 4   23. − 2.16   24. − 1.44

Solving Single Variable First Degree Equations in Three or More Steps - pgs. 134 - 136
1. 4   2. 5   3. 7   4. 9
5. 25/4   6. 15   7. 9   8. − 1
9. 29/12   10. 29/17   11. 4   12. − 3
13. 2   14. 3   15. 9   16. − 11/2
17. 2   18. − 5/4   19. 28   20. 45
21. − 15   22. 7   23. 4   24. 21
25. − 3/4   26. 2/3   27. 13/40   28. 1
29. 109/7   30. − 25

Solving Single Variable First Degree Inequalities in One or More Steps - pgs. 137 - 138
1. x > − 4   2. x < 20   3. x ≥⅔   4. x ≤96
5. x < − 3.7   6. x > 2.5   7. x ≤ − .1935 8. x ≥ − 1.86
9. x ≥3.5   10. x ≤ − 1.5   11. x > − 2/5   12. x < − 2.5
13. x ≤ − 1/9   14. x ≥ − 1   15. x > − 9   16. x < − 81
17. x > 4/3   18. x ≥ 2   19. x < 3   20. x ≤4.5
21. x > − 8   22. x ≥ 9.5   23. x ≤5.12   24. x < 6.08
25. x ≤ − 3/4   26. x ≥ 2/3

Solving Single Variable Second Degree Equations - p. 140
Note: The numbers in brackets {} are not ordered pairs but values of x satisfying the equation.
1. {− 2, − 9}   2. {1, 8}   3. {− 3, 9}   4. {4, 8}
5. {− 5/2, 3}   6. {− 7/5, 4}   7. {− 2, 1/3}   8. {− 1, 8}

Word Problems- p. 143
1. 5   2. 8, 15   3. 9, 11   4. 11, 16
5. 8, 47   6. 2 (reject − 9/4)

Consecutive Integer Problems - p. 145
1. 13, 14, 15   2. 6, 7, 8
3. $x^2 + 2x = x^2 + 2x$ 4. $x^2 + 3x = x^2 + 3x$

Motion Problems - pgs. 149 - 150
1. t = 5 hrs.   2. t = 2 hrs.   3. t = 4 hrs.   4. t = 5 hrs.
5. t = 2 hrs. 40 min   6. t = 6 hrs., 14 − t = 8 hrs.

Mixture Problems – p. 151
1. 12.5 lbs. Bermuda seed, 37.5 lbs. Kentucky Bluegrass seed.
2. 62.5 lbs. mixture (x = 12.5 lbs. Darjeeling tea)
3. 15 lbs. spinach (10 + x = 25 lbs. of mixture)

The Pythagorean Theorem - p. 154
1. b = 40   2. c = 10   3. a = 9   4. c = 25
5. b = 15   6. c = √34   7. c = √58   8. b = 2√10
9. a = 2√14   10. a = 4   11. c = 2√5   12. b = 2√5
13. 90√2   14. 24 ft.

Ratio and Proportion - p. 157
1. 8   2. 16   3. 12   4. 5
5. 24   6. 2   7. x = 10, x = 15   8. x = – 6, x = 12
9. x = – 8, x = 15   10. x = – 8, x = 20   11. 30 ft.   12. 57, 76

Percent – pgs.160, 162
1. 70   2. 13   3. 30   4. 546
5. 66 ⅔ % (≈ 66.7%)   6. 75%   7. 40%   8. 110%
9. 362.5   10. 200   11. 37.5   12. 115
13. $64.95   14. $217   15. $102.08   16. $750
17. $23,700

Solving Single Variable Radical Equations – pgs. 163 - 164
1. 121   2. 52   3. 67   4. 14
5. 1,094   6. 6/5   7. 85   8. 25
9. 24   10. 9   11. 20   12. – 4

Solving Fractional Equations - p. 165
1. 1/3   2. 1/5   3. 4   4. – 4   5. 4

APPLICATIONS IN TWO VARIABLES: X AND Y AS CHANGING VALUES
Expressing One Variable in Terms of the Other(s) - p. 167
1. $l = \dfrac{A}{w}$ 2. $r = \dfrac{C}{2\pi}$ 3. $b_1 = \dfrac{2A}{h} - b_2$

The General Form of a Linear Equation - p. 173

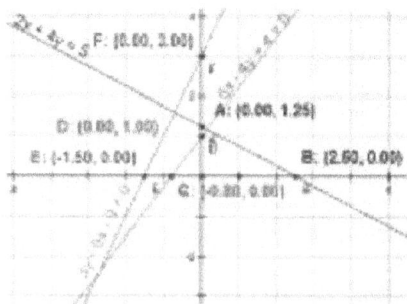

216

The Slope Formula - p. 176
1.   4/3   2.   2/3   3.   5/3

The Slope-Intercept Form of a Linear Equation - p. 179
1. $m = {}^3/_2, b = -3$ 2. $m = {}^1/_2, b = -2$ 3. $m = {}^2/_3, b = {}^4/_3$
4. $y = {}^2/_3 x - 3$ 5. $y = -{}^3/_4 x - 2$ 6. $y = {}^4/_3 x + 4$

Graphing Lines of the Form y = b - p. 180   and   Graphing Lines of the Form x = a - p. 181

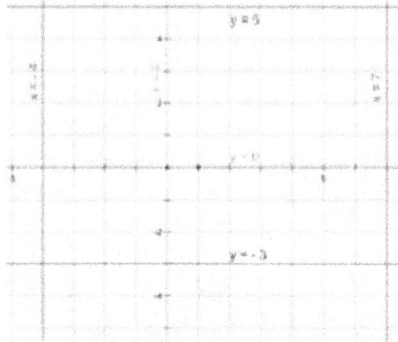

Writing the Equation of a Line Given the Slope and One Point - p. 183
1. $y = 3x - 27$ 2. $y = -2x - 4$ 3. $y = {}^2/_3 x - 2$ 4. $y = {}^1/_4 x + 2$
5. $y = 5$

Writing the Equation of a Line Given Two Points - p. 184
1. $y = {}^1/_2 x + 2$ 2. $y = -x + 2$ 3. $y = {}^2/_3 x - {}^4/_3$
4. $y = 4$ 5. $y = x - 5$

Solving Systems of Linear Equations in Two Variables - p. 186
1. $(2, -1)$ 2. $(3, -2)$ 3. $(-2, 1)$ 4. $(2, 4)$
5. $(4, 3)$ 6. $(9, 4)$

Graphing Systems of Linear Inequalities in Two Variables - p. 188

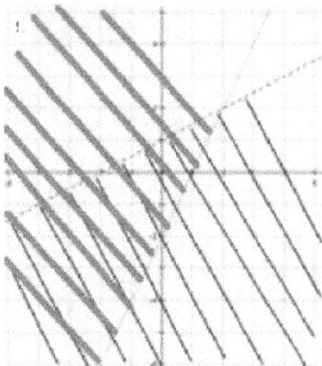

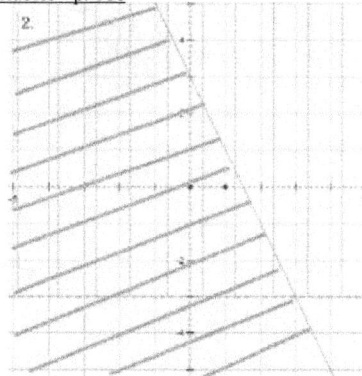

217

Quadratic Equations and the Parabola - p. 194
1. $(4, 0), (6, 0), (0, 24), (5, -1)$
2. $(1, 0), (2, 0), (0, 2), (3/2, -1/4)$
3. $(-1, 0), (4, 0), (0, -4), (3/2, -25/4)$

4, 5.

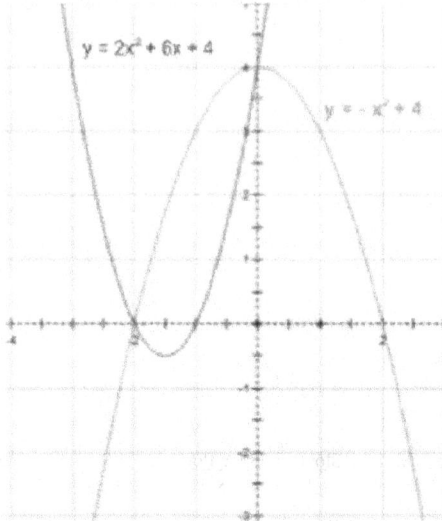

$y = 2x^2 + 6x + 4$

$y = -x^2 + 4$

Completing the Square - p.196

1. $x_1 = 1, x_2 = 3$
2. $x_{1,2} = 2 \pm \sqrt{5}$
3. $x_{1,2} = 4 \pm \sqrt{2}$
4. $x_{1,2} = 5 \pm \sqrt{5}$
5. $x_1 = 3, x_2 = 4$

The Quadratic Formula – p. 200

1. $x_1 = \tfrac{3}{4}, x_2 = -\tfrac{1}{4}$
2. $x_{1,2} = 2 \pm \sqrt{7}$
3. $x_{1,2} = 1 \pm \sqrt{2}$
4. $x_{1,2} = 1 \pm \sqrt{3}$
5. $x_{1,2} = 2 \pm \sqrt{3}$

www.ingramcontent.com/pod-product-compliance
Lightning Source LLC
Chambersburg PA
CBHW060014210326
41520CB00009B/880